Discovering

Ku-ring-gai Chase National Park

On Foot

by

Alan Fairley

envirobook

Published by Envirobook, The Gloucester Walk,
88 Cumberland Street, Sydney NSW 2000.
Telephone: (02) 247 6036

National Library Cataloguing in Publication Data:

Fairley, Alan.
Discovering Ku-ring-gai Chase National Park on foot.

ISBN 0 85881 130 8.

1. Hiking – New South Wales – Ku-ring-gai Chase National Park – Guidebooks. 2. Walking – New South Wales – Ku-ring-gai Chase National Park – Guidebooks. 3. Ku-ring-gai Chase National Park (N.S.W.) – Guidebooks. I. Title.

919.4410463

Designed by Jennifer Gormley, Envirobook, Sydney
and set in 10/12 Palatino.
Printed by Southwood Press, Marrickville, NSW
on 100% recycled paper using soy based ink.

Contents

Introduction

For 100 years, Ku-ring-gai Chase National Park has been inspiring visitors with its magical combination of rugged scenery, panoramic views, extensive waterways and its wealth of wildflowers and bird life. Located as it is abutting Sydney's northern suburbs and with easy access by road and public transport, it is not surprising that Ku-ring-gai Chase continues to attract thousands of visitors each year.

This book is designed to encourage some of these visitors to discover the beauty and variety away from the roads and picnic areas. For those who already know the joys of walking in the Chase, this book may open new routes to explore.

Descriptions of walks include distances, average time taken, gradient of the track, degree of difficulty and appropriate maps. These enable the walker to set out with confidence, knowing in advance the nature of the track. Information about plant species, geology, animal life, Aboriginal sites and European history is included so the walker can obtain the most from the outing.

Twenty-nine self-guiding walks are described in this book, ranging from short strolls of 15 minutes to longer walks of over four hours. The routes for seven more strenuous walks are also suggested. Park facilities, camping and accommodation information and details of public transport are all described in the General Introduction.

The more you explore the park, the more you will enjoy it – and the best way to explore is on foot.

The bushfires of January 1994 burnt much of Lambert Peninsula and West Head, reducing its varied plant communities to ash and blackened trunks, and taking a tragic toll on wildlife. The vegetation is slowly recovering but in some areas it will be a number of years before the wildflowers return to their former glory.

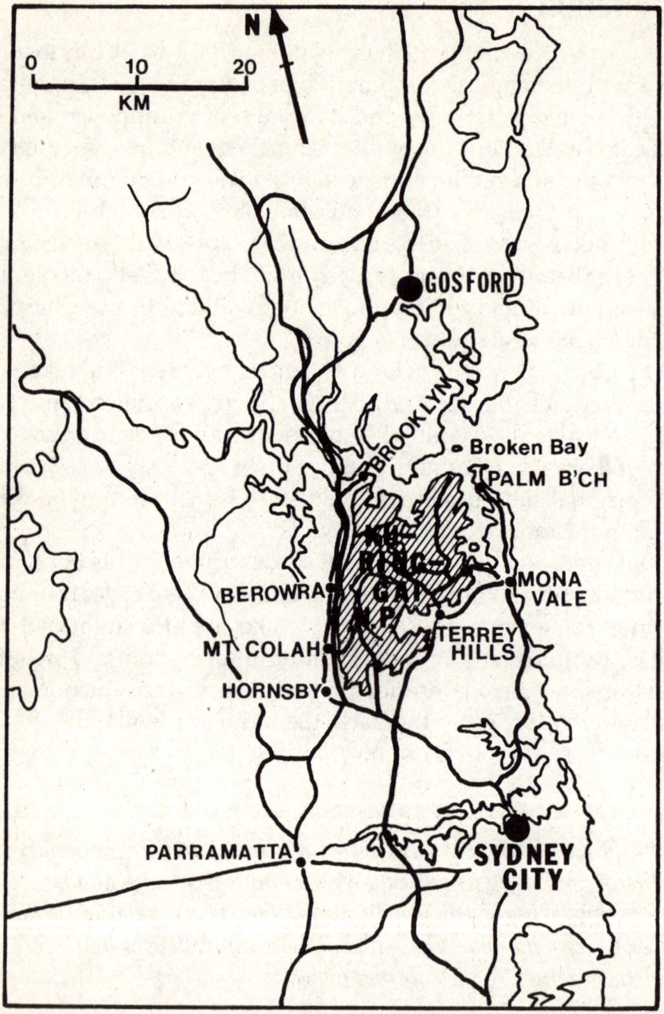

Location Map

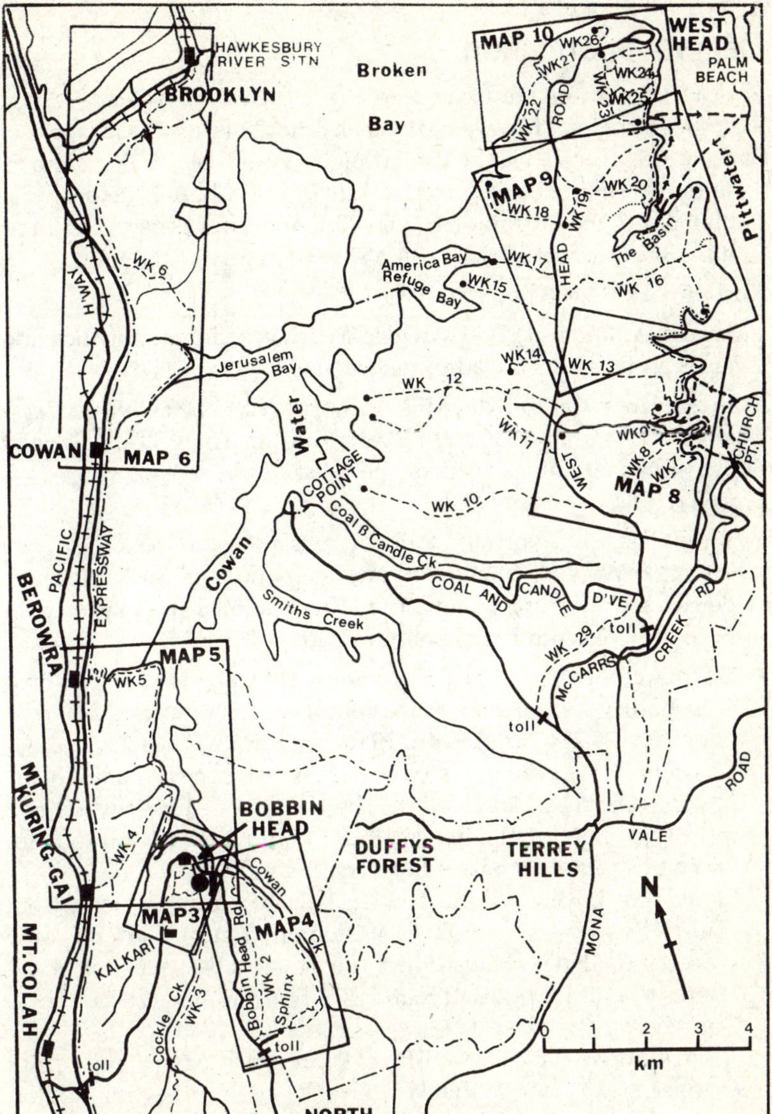

Ku-ring-gai Chase National Park

General Information

Park Headquarters are located on Ku-ring-gai Chase Road, 1.3 km from Bobbin Head towards Mt Colah, and 30 km by road north of Sydney's city centre via the Pacific Highway and North Turramurra.

Enquires should be directed to : Ku-ring-gai Chase National Park, Bobbin Head, via Turramurra. 2074. Telephone for general enquires: 02 457 9322 (Monday–Friday), 02 457 9310 (weekends).

Size of park. 14,838 ha.

Entry fee. A charge of $7.50 is made for each vehicle entering the park. Cyclists and walkers are admitted free.

Map reference. Central Mapping Authority (CMA), *Ku-ring-gai Chase National Park,* (scale 1: 40,000), available from The Wildlife Shop at Bobbin Head and other outdoor specialist stores.

Public transport

- **Train.** The Main Northern Railway runs along the western edge of the park. Walking tracks lead into the park from Mt Kuring-gai, Berowra, Cowan and Hawkesbury River stations. It is 8 km by road from Mt Colah Station to Bobbin Head.

- **Ferry.** Access to the park along western Pittwater is by ferry. There is an hourly ferry service from Palm Beach (Pittwater side): telephone 02 918 2747 or 02 974 5235 for ferry times. The ferry route is Coasters Retreat (Bennetts Wharf)–Bonnie Doon–The Basin–Currawong–Great Mackerel Beach. There is also a 24 hour water taxi from Palm Beach (telephone 018 238 190). The Wynyard to Palm Beach bus route 190 passes the ferry wharf.

  Elvina Bay, Lovett Bay and Towlers Bay are reached by ferry from Church Point (telephone 02 99 3492 for ferry times). The ferry route is Scotland Island – Halls Wharf (for Towlers Bay) – Lovett Bay – Elvina Bay. Bus routes 155 and 157 from Manly Wharf run to Church Point.

  A ferry departs at 11.00 am from Palm Beach wharf (daily from October to May) for Bobbin Head, via Patonga and Cowan Water. It leaves Bobbin Head for the return trip at 1.45 pm, reaching Palm

Beach at 3.30 pm (telephone 02 918 2747 for details).

- **Bus.** There is a regular bus service from Turramurra Station to the park entry gates at North Turramurra, 4 km from Bobbin Head (route 577 Shorelink, telephone 02 457 8888). This service goes all the way to Bobbin Head as follows :
  Depart Turramurra (weekdays) : 9.18 am, 1.14 pm.
  Depart Turramurra (Saturday) : 10.14 am, 1.14 pm, 4.14 pm.
  Depart Turramurra (Sunday) : 10.10 am, 12.10 pm, 3.10 pm, 5.10 pm.
  From Bobbin Head (weekdays): 1.34 pm.
  From Bobbin Head (Saturday) : 1.37 pm, 4.37 pm.
  From Bobbin Head (Sunday) : 1.28 pm, 3.28 pm, 5.28 pm.

Road access. The majority of visitors enter the park by road, heading for either Bobbin Head or West Head. **Bobbin Head** is connected by road from the Pacific Highway and Newcastle Freeway at Mt Colah and from North Turramurra by Bobbin Head Road. **West Head** and **Lambert Peninsula** are reached from the northern suburbs. West Head Road leaves McCarrs Creek Road between Church Point and Terrey Hills (32 km from the city), enters the park and passes along the length of the peninsula for 13 km to West Head Lookout. Numerous walking tracks lead off from West Head Road.

Coal and Candle Drive is a pleasant 12 km circuit , with a 8 km (return) branch to Cottage Point. This Drive takes you past the picnic area at Illawong Bay and the marina and boat hire facilities at Akuna Bay.

Guided walks. Chase Alive volunteers conduct regular activities throughout the year in and around Ku-ring-gai Chase. For bookings and programme telephone 02 457 9853.

Wildflower Garden. On the edge of the park at St Ives is the Ku-ring-gai Wildflower Garden operated by Ku-ring-gai Municipal Council. It is a good place to learn about the native plants of the area. Access is from Mona Vale Road. An admission fee is charged. The Garden is open daily from 7.00 am to 4.00 pm. Phone 02 498 0888 for details.

Facilities and Services

Bobbin Head. Extensive grassy picnic areas, shelters, barbecue facilities, toilets, drinking water, children's playgrounds, kiosk, canoe launch site, Field Studies Centre. Boats may be hired at Halvorsen's. The Wildlife Shop in the old Bobbin Inn sells cards, maps, books and gifts (open seven days).

Kalkari Visitor Centre. Information desk staffed by volunteers of Chase Alive. Exhibits depicting the natural history of the park; audio-visual presentations. The Bush Discovery Track is an easy 30 minute loop walk, suitable for the elderly and disabled, with signs to draw attention to interesting aspects of the local environment. Grey kangaroos and emus may be seen within a large fenced area. The Centre is open seven days a week (except Christmas Day) from 9.00 am to 4.15 pm.

Appletree Bay. Picnic areas, toilets, public phone, boat ramp and kiosk at weekends.

The Basin. Camping and picnic facilities, toilets, public phone.

West Head Lookout. Picnic areas, toilets.

Resolute (Garigal) Picnic Area. 800 m before West Head Lookout. Picnic tables, barbecue facilities, information boards, water, public telephone.

Akuna Bay. On Coal and Candle Creek. Kiosk, toilets, public phone, boat ramp, restaurant, cafe, mini mart. Boat hire from privately operated marina.

Cottage Point. Kiosk, general store, restaurant, boat hire.

When To Visit

Any time is a good time to walk in the park. Sunny autumn and winter days are ideal for longer hikes. August to October brings the main burst of wildflowers. In summer, walkers will find plenty of cool havens along Cowan Creek and Pittwater, with beaches and pools in which to cool off. Bobbin Head and other picnic grounds with road access are especially crowded over Christmas and Easter holidays, but even then walkers will find many isolated and peaceful spots. The park may be closed at times of high fire danger.

Camping and Accommodation

Camping is not permitted in the park except at the serviced camping ground at The Basin. Access to The Basin is by ferry from Palm Beach. For bookings and information, telephone 02 974 4036.

There is a YHA hostel on the southern side of Towler Bay on Pittwater. Access is by ferry from Church Point or by foot from West Head Road. For bookings, telephone 02 992 196.

Currawong Beach on Western Pittwater is a 23 ha bushland property owned by the Labor Council of NSW. It has self-contained cottages for hire to the public at reasonable rates. Phone 02 974 4141 for details. Access is by ferry from Palm Beach.

Looking after the park

The National Parks and Wildlife Service seeks your co-operation in observing a few common-sense regulations designed to protect you and the park.

- **Litter.** Carry out all rubbish. Do not bury or leave cans, paper, plastic, bottles or other rubbish.
- **Pets and firearms.** Leave them at home.
- **Fire.** Camp fires are not permitted. Fires may only be lit in authorised fireplaces within picnic areas.
- **Natural features.** All animals, plants, rocks and soil are protected and must not be disturbed.
- **Noise.** The park is a place for nature – leave radios and cassette players at home and keep other noise to a minimum.

Walking Hazards

If care and common sense are used, few dangers will confront the walker in Ku-ring-gai Chase. Plan your outing. Take a map. Take sufficient food and fresh water. Solid but comfortable walking shoes are desirable. In hot weather, wear light loose clothing and a hat; be prepared for late afternoon storms. Don't forget the sunblock, insect repellent and anti-itch cream.

Specific hazards include :

- **Ticks.** These are found in drier areas. They may drop onto you and burrow into your skin without you knowing. An itch is usually the first indication you have a tick. Tweezers are needed to pull it out. If the head breaks off and remains in the skin, infection could occur.
- **Leeches.** Uncommon in the park; confined to the wetter gullies. A liberal spray of insect repellent on socks and shoes will help keep them away. If you find a leech on your skin, simply pull it off; salt or a burning match will make it release its grip quickly. Most leech bites itch for a few days; a few people have allergic reactions to the bites.
- **Snakes.** There are venemous snakes in the park, but they are seldom encountered. If you see a snake, wait for it to move off or walk around it. Few snakes are aggressive unless provoked. If bitten, keep calm and still, apply a pressure bandage to the bitten area and send someone for help.
- **Prickles.** As tracks are sometimes overgrown with prickly understorey plants, long pants are advisable.
- **Bushfire.** Avoid walking in weather in which fire is likely to occur. At times of peak fire danger, the park will be closed. If you see smoke, assess your position and head for the water or the main road. Remember that radiation is the principal danger, so try to cover all exposed skin and shield yourself from the radiated heat.

Park History

Ku-ring-gai Chase National Park was established in 1894, largely due to the efforts of one man, Eccleston du Faur, who worked tirelessly to convince a reluctant government that a park was necessary 'to prevent the reckless destruction of native flowers' and because of the 'rapidly increasing neighbourhood' (letter from du Faur to Henry Copeland, 1892). Its original area was 35,000 acres (14,175 ha).

Du Faur was a surveyor and engineer, a Latin and Greek scholar, a supporter of the arts and exploration, and had an interest in flora and fauna. He lived at Turramurra and was keen to see his beautiful 'back yard' protected. In 1892 he even invited the governor, the Earl of Jersey, and his son to meet him for a picnic and boat trip on Cowan Creek. Because the governor got lost on the way, the boat trip had to be cancelled but the picnic took place on top of the plateau.

Prior to the establishment of the park, there was little attempt to settle the area because of its sandy soils and difficult terrain. Access was mostly by water. Sometime before 1856, Duffy's Wharf was built on the upper reaches of Cowan Creek to transport cut logs downstream. Cottage Point had been granted to James Terry in 1885 and there was a boatshed in Waratah Bay in the early 1890s.

The existence of richer soils along Western Pittwater meant that there was an early interest in that area's farming potential. James Larmer was instructed to survey the lands along Pittwater in 1832; he noted that some farms had already been established and that there were huts on Great Mackerel Beach and at The Basin. In 1866 Commander Sidney formally named the bays and headlands of West Head. Commodore Lambert, captain of HMS *Challenger*, also named a number of bays on his inspections of the coast between 1866 and 1870. The land around West Head Lookout was not originally included in the park as it had been a land grant to the explorer William Lawson. It was not until 1951 that the 640 acres (260 ha) of West Head became part of the national park.

The park has seen a number of changes since its formation. Roads were constructed to Bobbin Head from Turramurra between 1894 and

1901, from Mt Colah to Bobbin Head in 1903 and an improved road and bridge at Bobbin Head were completed in 1929. A dirt track existed along the length of West Head Peninsula from the 1920s, but the current West Head Road was not redesigned, surfaced and opened for traffic until 1964.

A number of schemes for the park never came to fruition. These included plans for rifle ranges, golf links and a marine drive from Bobbin Head to Brooklyn. In 1899, it was even proposed to establish the new capital of Australia between Cowan Creek and Smiths Creek. In 1929, a prospectus was issued to raise funds for the development of West Head as a country estate with hotel, casino, golf course and public housing lots; fortunately this came to nothing because of the Great Depression.

The park today stands as a living testimony to the foresight of Eccleston du Faur and to those administrators who resisted proposals which would have compromised its natural values and beauty.

Geology and Landforms

Ku-ring-gai Chase is part of a massive block of sandstone, tilting upwards to the north, known as Hornsby Plateau. The park itself consists of a plateau with an elevation of about 200 m above sea level, reaching a maximum of 246 m at Willunga Trig off West Head Road. The V-shaped flooded valley of Cowan Water cuts into this plateau, separating the park into two sections. Creeks flowing from the plateau area form deep and rugged gullies, flanked by steep tree-covered slopes and separated by narrow ridges. The waters of Cowan Creek, Broken Bay and Pittwater almost encircle West Head Peninsula.

A feature of the park's landscape is the 'flooded' or 'drowned' valleys. During the last one million years, when water was locked in the ice caps during the various ice ages, the sea level was lower and the streams eroded valleys much deeper than is now possible. When the ice caps melted, this water was unlocked and sea levels rose, flooding the valleys. Broken Bay, Cowan Water and Pittwater were all formed in this manner about 10,000 years ago, as were the deeply incised inlets such as Jerusalem Bay, Coal and Candle Creek, Smiths Creek, Cowan Creek, The Basin and McCarrs Creek.

Hawkesbury Sandstone is the dominant rock type of the park and accounts for both the topography and the characteristic vegetation of the sandy soils. Sandstone cliff lines, rocky knolls, rock outcrops and weather-worn sandstone caves are common. Aborigines used the flat sandstone outcrops as sites for rock engravings. The formation known as 'tessellated pavement' often occurs; it features a regular chequered pattern across the surface of the rock along fault lines or lines of weakness. Streams also follow these fault lines, cutting down through the bedrock or, as at America Bay and Lovett Bay, falling as a waterfall over a harder layer of the outcropping sandstone.

Other rock types are not common in the park. Wianamatta Shale does overlie the sandstone on a few high points. Eroded remnants of volcanic dykes also occur, such as near Resolute Picnic Area at West Head and at Campbells Crater near Cowan. These produce a richer soil marked by the presence of tall open forests. Soils derived from the Narrabeen Group are found along the lower slopes and shoreline of Western Pittwater, especially around Elvina Bay, Lovett Bay and Towlers Bay. Here the richer shales and sandstones make the vegetation distinctly different from that found on top of the ridges and give rise to fine spotted gum forests with forest oak as a secondary layer.

Plant Communities

COMMUNITY	SOILS	OCCURRENCE	TYPICAL PLANT SPECIES
HEATH	Hawkesbury Sandstone: shallow skeletal soils	Ridge tops (shallow rocky soils)	*Banksia oblongifolia*, darwinia , dwarf apple, grass tree, sedges. Many representatives of Proteaceae, Epacridaceae and Fabaceae.
CLOSED DRY SHRUBLAND	Hawkesbury Sandstone	Ridge tops, upper slopes and drainage lines (deeper sandy soils)	Heath banksia, old man banksia, black she-oak, hakea spp, tea-trees, pinnate boronia.
WET HEATH/ SEDGELAND (hanging swamp)	Alluvium/ impervious clay and rock	Upper reaches of streams	Swamp banksia, pink swamp heath, prickly tea-tree, coral fern, dog rose, gahnia and other sedges.
WOODLAND	Hawkesbury Sandstone	Drier plateau tops and upper slopes (sandy soils)	Scribbly gum, small red bloodwood, dwarf apple, old man banksia, grass tree, hakea spp; rich shrub understorey.
OPEN FOREST	Hawkesbury Sandstone	Gully slopes and valleys	Red bloodwood, smooth-barked apple, Sydney peppermint, Christmas bush, grass tree; rich shrub understorey.
TALL OPEN FOREST	Narrabeen Shales	Lower slopes, West Pittwater	Spotted gum, grey ironbark, forest oak, burrawang, boronia *(B.fraseri, B. mollis),Astrotricha floccosa.*
WARM TEMPERATE RAINFOREST	Sandstone and shales	Deeper protected gullies	Coachwood, cabbage tree palm, kanooka, grey myrtle, lillypilly.
MANGROVE	Alluvial, tidal mud flats	Shoreline of bays and inlets	Grey mangrove, river mangrove; swamp oak on margins.

Floral Calendar

JANUARY	Many eucalypts in flower (eg, Sydney peppermint, grey gum). Dwarf apple, old man banksia. many small understorey plants in flower.
FEBRUARY	Banksias (*B. integrifolia, B. marginata, B. robur*). Bark changes of smooth-barked apple.
MARCH	Red bloodwood, *Banksia spinulosa, Banksia oblongifolia, Crowea saligna, Lambertia formosa, Epacris pulchella.* Numerous fungi.
APRIL	Banksias, crowea, woollsia, grey spider-flower, she-oaks, fuchsia heath, some early wattles.
MAY	Banksias (*B. ericifolia, B. oblongifolia, B. marginata*), wattles (*A.suaveolens, A. ulicifolia*), pink tea-tree, early pea flowers, *Epacris pulchella,* grass trees (in flower after summer fires).
JUNE	Grevilleas (red, pink, grey), early wattles, red styphelia, darwinia, dog rose, fuchsia heath, woollsia, *Banksia ericifolia, Banksia integrifolia, Leucopogon microphyllus, Hibbertia* spp.
JULY	Boronias (*B. ledifolia, B. pinnata, B. mollis*), red grevillea, grey grevillea, flannel flower, numerous wattles.
AUGUST	Beginning of main flowering in heath and shrubland. Boronias, pea flowers, ground orchids, epacrids.
SEPTEMBER	Main flowering in heath and shrubland. Many species, dominated by pea flowers, grevilleas and members of Rutaceae.
OCTOBER	Many plants still in flower. Tea-trees, drumsticks, black-eyed Susan, cabbage tree palm.
NOVEMBER	Summer flowering begins. Smooth-barked apple, lillypilly, phyllota, trigger plant, woolly star-hair.
DECEMBER	Dwarf apple, Christmas bush, Christmas bell, guinea flowers, flannel flower, needle bush.

Animal Life

Walkers within the park cannot expect to see many mammals as most are shy, feed mainly at night and keep well away from areas used by humans. In addition, the bushfires of January 1994 drastically reduced animal numbers, especially on the West Head Peninsula.

The swamp wallaby is the largest animal in the park. It may be seen occasionally on a quiet day around Resolute Picnic Area at West Head. The pygmy possum feeds on nectar and insects on the heath. Long-nosed bandicoot, brown antechinus (marsupial mouse), ring-tailed possum and brush-tailed possum are all residents of woodland and open forest. The echidna is sometimes observed during daylight hours. Only one small colony of koalas survive in the park; their forest habitat was saved from the 1994 fires.

Over 160 species of birds have been recorded for the Chase. Honeyeaters are a feature of the heaths and shrubland and include the white-eared, white-cheeked, New Holland honeyeater, red wattlebird and eastern spinebill. In forested areas, the brown thornbill, grey thrush, grey fantail, white-throated treecreeper and pardalote are often observed. The white-faced heron is common on mangrove flats, whipbirds and lyrebirds in gullies, and noisy miners, kookaburras, currawongs and magpies around picnic areas.

A wide range of reptiles, including skinks, geckoes, dragons, venomous and non-venomous snakes, are common in summer months, but most are well camouflaged. Walkers will occasionally come across a lace monitor (goanna) as it scrambles up a tree, a copper-tailed skink as it scurries for cover beneath a rock, a jacky lizard in the leaf litter or a red-bellied black snake as it suns itself on a rock, but generally reptiles appear to be just as wary of humans as we are of them.

Aboriginal Pre-history

There are many indications of past Aboriginal habitation throughout Ku-ring-gai Chase. Rock engravings are relatively common on expanses of flat sandstone. Some charcoal and ochre paintings, including hand stencils, have been found. Axe grinding and tool sharpening grooves often occur in the sandstone beside water, especially in small streams. Shell-heap middens are abundant along the shores of Cowan Creek. Many of the rock shelters show signs of Aboriginal occupation.

Examples of all these sites may be seen beside the walks described in this book. Rock engravings occur along West Head Road, beside Elvina, The Basin, America Bay and Resolute Tracks and on the old Bobbin Head Road. Red hand stencils occur in a small sandstone overhang near Resolute Picnic Area. Axe grinding grooves are widespread; the easiest to see are those along America Bay Track just 160 m in from the road. The foreshore tracks along Cowan Creek all cut

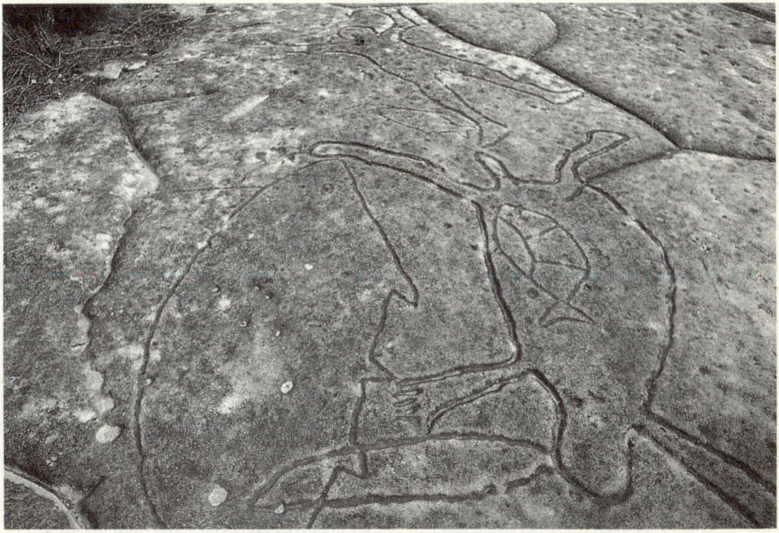

Aboriginal engravings – West Head

through midden heaps. A fine example of an Aboriginal rock shelter occupation site is on the lower section of the Resolute Track above Resolute Beach.

The Aboriginal people who lived north of Sydney spoke a language known as Guringai (or Kuringai). They may have lived in the area for 30,000–40,000 years, but physical evidence older than 8,000 years is difficult to find given that sea levels rose from about 17,000 years ago covering the lower coastal sites. As most of the area is near water, fish would have been an important part of their diet, caught by spears or by hook and line from bark canoes. Fish, eel and whales feature regularly in the engravings. Shellfish were collected by the women.

Thick layers of discarded shells (middens) are found along the shoreline, often near shelter caves. Traces of charcoal indicate fires were built on the midden sites and the presence of stone implements and waste flakes also suggest they were tool industry locations. Middens around The Basin were reported by early explorers to be 6–9 m thick. They were so extensive that there was a shell collecting industry in Western Pittwater until 1870. The shells were taken to be burnt for lime needed to supply the cement for an expanding Sydney.

The coastal Aborigines were hunters and gathers who exploited the varied food resources of the sea and land. Archaeological evidence from middens suggests that in addition to fish and molluscs, their diet also consisted of birds, bats, dolphins, wallabies, possums, lizards and smaller mammals. They would also have used the fruit, seeds and roots of native plants.

Rock engravings often show hunting scenes. These may have been stories of successful hunting trips or an attempt to use 'magic' to ensure a successful hunt. We can only guess the meanings of these engravings. No doubt many of the ancestral spirits depicted would have had a role in the initiation of young men, while others show popular tales about the adventures of Dreamtime beings.

The Guringai people had nearly all died within 50 years of the European arrival, so little information about them was recorded.

Bobbin Head and Northern Railway

1 BIRRAWANNA TRACK

Distance 4.3 km (round trip)

Time 2 hours

Features A good walk for those who have only limited time and yet wish to see the main features of the bushland around Bobbin Head. The open woodlands of the plateau have a fine variety of flowering plants and bird life is abundant. Scenic views over Cowan Creek.

This walk is along a well defined , well used track linking Kalkari Visitor Centre, Bobbin Head and the Park Headquarters/District Office in a circuit. It is graded easy to medium. From Kalkari you cross the plateau before descending 160 m to Bobbin Head. The return route is split into two 80 m ascents, the first out of Bobbin Head, the second up the slope from Park Headquarters. Both ascents are rather steep and demanding.
Although this description begins at Kalkari, it is just as easy to pick up the track at Bobbin Head or at Park Headquarters. A pleasant, easy alternative is to arrange a car pick-up at Bobbin Head and to walk one way down hill from Kalkari (30 minutes).

The starting point, Kalkari Visitor Centre, is located on Ku-ring-gai Chase Road, 4.2 km from the Pacific Highway and 3.2 km from Bobbin Head. Birrawanna Track begins at the northern end of the car park. It is signposted. At first it follows the wire fence surrounding the emu compound, then climbs gently to the ridge-top. The open woodland here is rich in flowering shrubs, such as spider flower, flannel flower, waratah, waxflower and various pea-flowers and epacrids.

After you cross the ridge-top service trail, you begin to descend towards Bobbin Head. This is a very old trail, used by horsemen long

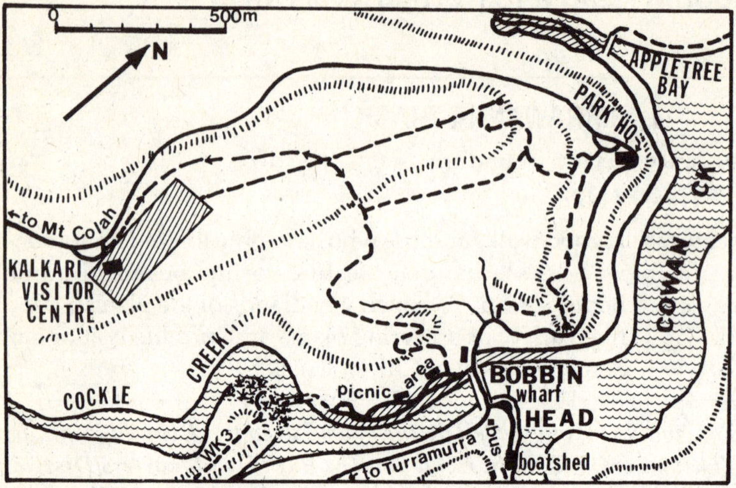

Kalkari–Bobbin Head, Birrawanna Track

before the main road was built. Some physical evidence of the old track still remains, such as stone gutters, hand-hewn cuttings, rock edging and a dump of cut stones. A track branches off to your left; this is an alternative route around the slope to Park Headquarters.

The vegetation changes on the more sheltered slopes. Beneath a canopy of grey gums, Sydney peppermints and forest oaks, there are slopes of bracken fern and the upright shrub *Astrotricha floccosa,* which has large leaves, conspicuous clumps of whitish flowers in late spring and soft woolly hairs covering its stems. Further down the slope, the track passes along a drier open ridge overlooking Cowan Creek, then descends steeply via some stairs, passes under a rock overhang, to emerge at the picnic area of Bobbin Head between the Field Studies Centre and Bobbin Inn.

Walk up the road past Bobbin Inn. The return track commences just around the first corner beyond the second gully, 150 m up from Bobbin Inn. Note the tall coachwood and lillypilly growing in the shaded, rocky gully. The track ascends steeply until it reaches a ridge. From the rocks at the end of the ridge to your right is a fine view over Cockle Creek and Bobbin Head. The track now takes a more gentle gradient along the slope parallel to the road. It is only about 15 minutes to the Park Headquarters. Here there is drinking water beside the car park.

The route is signposted from the car park. The track winds up the slope past some beautiful scribbly gums to a grassy flat where a youth hostel stood in the 1960s and 1970s. Little remains today except some foundation bricks and an old fireplace. Further up the slope, the track passes beneath grey gums and scribbly gums; red bloodwoods are common higher up. Near the top of the slope, the track divides. Take the right hand fork which ascends to the service trail on the ridge-top. Turn right and take a few minutes to rest and gaze out over the park and Cowan Creek from the end of the trail.

Return along the service trail. The vegetation here is typical of the dry open woodlands of sandy ridges throughout the park. *Banksia spinulosa* and *Epacris pulchella* flower in autumn. Tea-trees, conesticks, geebungs, pea-flowers, hakeas and numerous other species can be seen flowering in late winter and spring.

Turn right when the Kalkari–Bobbin Head track crosses the fire trail and return to your starting point.

2 NORTH TURRAMURRA – BOBBIN HEAD – SPHINX

Distance 9.6 km (round trip)

Time 3 hrs 15 mins

Features This circuit includes the historical features of the old Bobbin Head Road and Sphinx, a set of Aboriginal engravings, plateau, gully and estuarine vegetation, plentiful bird life and a pleasant stroll along the margins of upper Cowan Creek. Boat hire and a kiosk are available at Bobbin Head. You may wish to do this walk one way downhill to Bobbin Head, either via Old Bobbin Head

Road (4 km) or via the Sphinx Track and Cowan Creek (5.6 km), then return by bus.

Old Bobbin Head Road is a wide service trail, no longer used as a road. For 3 km it is flat and easy going; it then drops steeply in a series of bends to Bobbin Head. The return along Warrimoo Track and Cowan Creek is narrow but well defined, mostly easy but with a few short, sharp ups and downs. The ascent from the valley to the Sphinx (150 m over 1.2 km) is steep, especially the first section up steps from the creek.

From the park entrance, follow the main sealed road towards Bobbin Head for 100 m. A service trail leads to your right. Beyond the gate, turn left. You are now on the original Bobbin Head Road, begun in 1894 and completed in 1901. It was in use until a new road was opened in 1929, although at first traffic went down the new road and came up the old road. The old road was finally closed in May 1930.

The trail follows the ridge and is flat and easy going. The soils of the plateau are sandy with hard reddish laterite scattered over the surface. Hard-leaved shrubs dominate the understorey, especially needle-bush, honey flower, four species of banksia and epacrids. Scribbly gum (*Eucalyptus racemosa*) is the tallest tree ; dwarf apple is common.

After about 25 minutes walk, and beyond the high voltage power line, there is a small set of Aboriginal engravings on flat rocks to your left. Depicted are a male figure, a kangaroo/wallaby, an echidna, a goanna and a set of 'footprints' across the rock surface. Be careful to avoid walking on any of the engraved lines as this accelerates the erosion process.

Another kilometre further on, the trail rises to cross a knoll. Here the vegetation is taller and dominated by Sydney peppermint, smooth-barked angophora and old man banksia. On the northern slopes are red bloodwood and grey gum. A water tank is passed and from this point the trail descends in a series of zig-zags to Bobbin Head. The rock embankments and drains are testimony to the great effort which was put into the construction of this road.

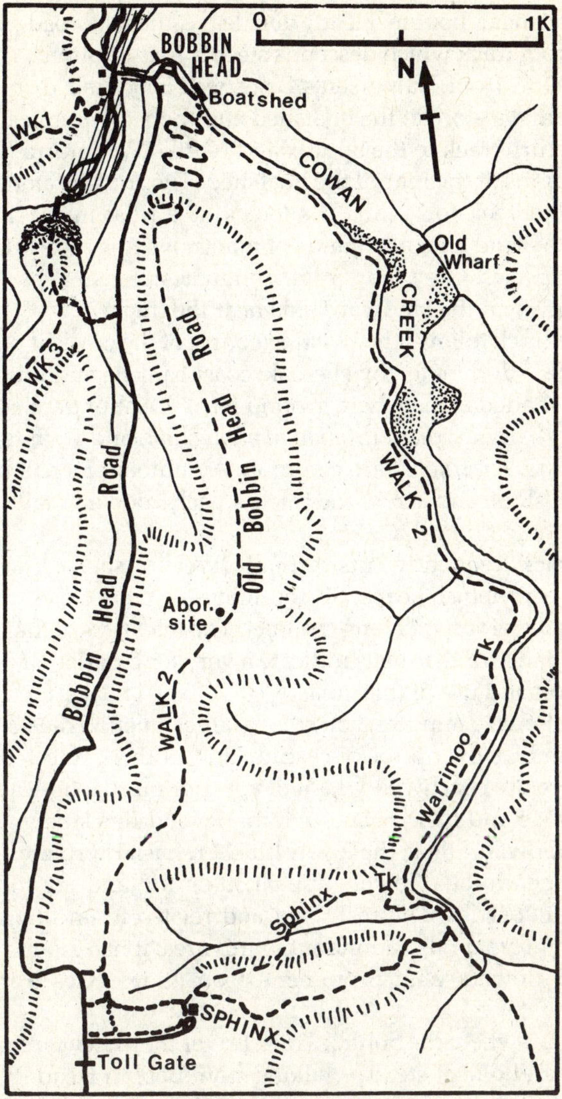

North Turramurra–Bobbin Head

As you near Bobbin Head, you leave the old road and take a marked foot track which descends steeply to the bottom. You emerge about 100 m from Halvorsen's Boatshed. Food and drinks may be purchased at a store at the boatshed and there are public toilets.

The return track to the Sphinx and North Turramurra (Warrimoo Track) begins at the rear of the boatshed. The route is along the edge of Cowan Creek for 4 km. The 'creek' is a tidal inlet with alluvial deposits of sand and silt. Stands of mangrove line the estuary. There are two species. Grey mangrove dominates; the smaller river mangrove grows on the landward side near the track.

As the track follows the twisting course of the creek, it passes from sunny areas dominated by she-oak, coast banksia and rough-barked apple to shaded areas where ground ferns are abundant and crowea (*C. saligna*), trigger plant (*Stylidium productum*) and a white-flowered grevillea (*G. linearifolia*) are common. A number of small creeks run down the slope and across the track. Coachwood and tall saw-sedge occur in these gullies.

The track frequently cuts through layers of shells. These are old middens, or rubbish heaps, of Aborigines who once fished and gathered molluscs for food along the inlet. The thickness of the layers and the abundance of the shells indicate a very long period of Aboriginal occupation and use of this area.

After about 40 minutes, you cross a larger creek by means of stones. Water tumbles over a rocky cascade and in this protected environment there are coachwood, kanooka, water gum, long-leaved lomatia, dog rose and banks of coral fern. Beyond this stream, along the banks of Cowan Creek, the lovely blueberry ash is prolific. Its masses of white pendulous flowers appear in spring and are followed by bright blue berries. The restricted and relatively uncommon shrub *Darwinia procera* is also abundant in this area; it has groups of four to eight red flowers with prominent protruding styles from July to October.

The point where the Sphinx Track leaves the valley is reached after just over an hour of steady walking from Bobbin Head. It is clearly marked and is just before a deeply incised side creek. This track ascends steeply via some stairs. It follows the creek gully, crosses the creek and continues up the slope. Near the top, it joins another track just behind the Sphinx. Turn right, then first left to reach the Sphinx and the tarred access road.

The Sphinx is a one-eighth scale miniature of the famous Egyptian monument. It was hewn from a block of sandstone by Private W. Shirley, who was a patient of nearby Lady Davidson's Home between 1925 and 1926. It took him eighteen months and was dedicated to the memory of his fallen comrades in the AIF.

From the Sphinx, it is less than a five minute walk back to the park gates.

3 GIBBERAGONG TRACK

Distance 8 km (return)

Time 3 hrs

Features An easy to moderate walk from Bobbin Head Picnic Area to Gibberagong Waterholes along the valley of Cockle (Spring Gully) Creek. Mangrove boardwalk, changing vegetation patterns and a variety of birdlife.

Note. This walk can be done one way downhill from the park boundary at the end of Grosvenor Street, Wahroonga . The distance to Bobbin Head is 6 km.

A narrow foot track, sometimes running along the creek bank, at other times along the slope above the creek. Some ups and downs but no long or difficult ascents. Formerly called Grosvenor Track. The NPWS plans to up-grade this track following the opening of the foot bridge over Cockle Creek.

From the bus stop, cross the road bridge at Bobbin Head and walk down the western side of Cockle Creek through the picnic area to the end of the road. A wide path takes you to a cable foot bridge over

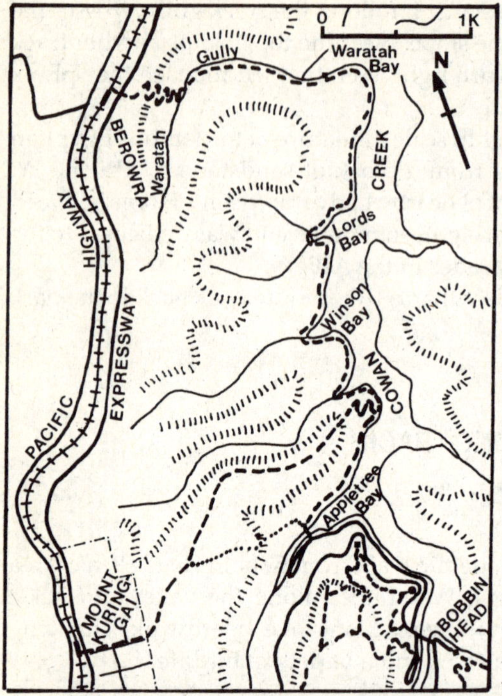

Cowan Creek

Cockle Creek and a boardwalk through the mangroves and fringing salt marsh. This bridge and walkway were constructed as a park centenary project (opened December 1994).

An alternative starting point is along the old track which begins 800 m from the road bridge up Bobbin Head Road towards Turramurra. This track drops immediately to cross a shaded gully, then skirts the slope to join the track from the mangrove walk on top of a saddle.

From the boardwalk, the track ascends to a small dry ridge. The scattered trees are chiefly grey gum, Sydney peppermint and smooth-barked apple. Shrubs, such as flaky-barked tea-tree, waxflower, conesticks and needle-bush, predominate. After passing the junction of the old track, you descend to the edge of the creek and walk behind the fringe of mangroves.

Two species of mangroves occur along the creek . Grey mangrove is the larger species, river mangrove is the smaller and occurs on the landward side. River mangrove also occurs further upstream than grey mangrove as it can survive in less salty water. Swamp oak grows on the margins of the tidal area.

Riverside flats have been invaded by exotic plants. At one point, there is a stand of the garden iris, *Watsonia* ; elsewhere there is castor oil plant, privet, balloon vine and camphor laurel. The native vegetation of the sheltered slopes is very varied. It includes narrow-leaf geebung, forest oak, silver banksia, Christmas bush, *Pultenaea flexilis*, *Hibbertia bracteata*, *Dampiera purpurea*, *Crowea saligna* and lots of grass trees. The rocky outcrops often have banks of ferns growing in damp clefts.

The valley is a good bird spotting location, especially in the early morning. You could expect to see yellow robins, scrub-wrens, thornbills, pardalotes, rock warblers, kingfishers, crimson rosellas, cockatoos, black ducks and pied cormorants, and hear the call of lyrebirds and whipbirds.

The upper tidal limit is marked by a jumble of large, grey, lichen-covered rocks in the stream. This area is surrounded by tall lillypilly, grey myrtle, rough-barked apple and the wattle *Acacia parramattensis*.

The track now climbs steeply up the sheltered slope through ferny banks and stands of *Boronia fraseri*. Further upstream, the slopes are more open and sunny. You cross a small stream, skirt a steep slope from where you look down on the creek, drop back to the creek's edge, then follow the upper bank as it curves around. A jelly-bean shaped pool lies in a bend of the creek; sandy banks are conspicuous on the opposite side.

Gibberagong Waterholes are just three minutes further on. Here, at the junction of Spring Gully Creek from Hornsby and Lovers Jump Creek from North Turramurra, there is a large pool formed by rocks which block the flow of the combined creeks.

The track continues over Lovers Jump Creek and steeply climbs the slope to Grosvenor Service Trail, but unless you want to continue on to Wahroonga Station (4.6 km further), return the way you have come.

4 MT. KURING-GAI – APPLETREE BAY – BEROWRA

Distance 11.3 km (station to station)

Time 3 hrs 40 mins

Features A combination of scenic beauty, varied plant life and some Aboriginal and early European historical sites. The core of this walk is along the edge of beautiful Cowan Creek below steep wooded slopes. As the walk begins and ends at railway stations, it is particularly suited for those who do not have a car.

A well-defined track descends gradually 210 m from the ridge-top at Mt. Kuring-gai to Cowan Creek. The track skirting Cowan Creek is narrow and sometimes eroded but easy to follow and mostly flat. The most difficult part of the walk is the steep ascent in a series of zig-zags back to the ridge and Berowra Station from Waratah Bay, a constant ascent taking 50–60 minutes. Medium difficulty.

From Mt Kuring-gai station, cross the footbridge over the expressway and walk down Harwood Avenue for 300 m to the park boundary and the start of the track.

At first the route is through open woodland. The scattered trees are mostly scribbly gum, red bloodwood, Sydney peppermint and old man banksia. The understorey depends on the frequency of fire but is rich in species. As is usual in such communities, the dominant plant family is Proteaceae, containing such species as long-leaf cone-seeds, honey flower, pink grevillea, needle-bush, crinkle-bush, gee-bungs and silver banksia. Members of the pea family are represented by *Bossiaea, Aotus, Gompholobium, Daviesia* and *Pultenaea.* Sydney boronia, waxflower, woollsia and a number of wattles are also present.

In less than ten minutes walk from the park gate, there is a junction in the track. Keep left. After another three minutes you come to the end of the ridge and the track descends beside a large rock outcrop.

The descent is gradual until near Cowan Creek where the track makes a series of bends down the steep slope. Note the change in vegetation on this sheltered slope. The dominant tree is forest oak, the ground is covered with ferns and there are scattered shrubs and small trees, including blueberry ash, scentless rosewood, grass trees and woolly star-hair.

Turn right on the lower track and follow it along to Appletree Bay. You pass the site of Woodnutt's Boatshed, now marked by old stone walls and exotic vegetation. The track cuts through numerous shell middens. Such middens are a common feature of the margins of Cowan Creek. They are often very thick and are evidence of a long history of Aboriginal occupation in the area. The middens are often in or near rock overhangs which provided shelter for wandering bands of Aborigines.

Appletree Bay is a popular picnic and boating spot. There is a small kiosk (open only at peak times), toilets, drinking water and public telephone. It is 1.4 km by road to Bobbin Head where you can catch a bus to Turramurra (infrequent service) or join the Sphinx/North Turramurra track (Walk 2).

The track from Appletree Bay to Waratah Bay winds along the edge of Cowan Creek. It takes about 90 minutes of steady walking to cover the 4.8 km, although you will probably take longer as there are many picturesque spots along the way. As the track follows the curves of the stream and around three bays, it passes through both open and shaded zones so the vegetation is constantly changing. Huge rough-barked apples and grey gums grow along the shoreline.

Creeks flow into each of the three bays. They fall steeply from the plateau in shaded gullies which are full of large boulders. The rocks and trees are covered with lichen and rock-felt fern. Large coachwood and kanooka thrive; grey myrtle and lillypilly are common. In 1911, several Aboriginal bones and stone axes were found in a cave at Winson Bay (the second bay you come to) and there are large midden heaps on the sunny side of Lords Bay (the third bay).

You cross two creeks at the head of Lords Bay, one of which falls over a sandstone cliff. It is about 35 minutes from here around to

Waratah Bay. This track was completed in 1910 at a cost of £50 ($100).

Waratah Bay is the site of the old Windybank's Boatshed, constructed in the 1890s and destroyed by bushfire in August 1972. Today all that remains are some foundations in the water, rusting metal hulks stranded in the sand, a few exotic plants, some water pipes and clearings beside the bay. (For more details of Waratah Bay see Walk 5).

It is 2.5 km (50–60 minutes) back to Berowra station, at first up a broad track along Waratah Gully, then by a zig-zag track up the steep slope. Near the top of the ridge, the route turns sharp right away from the old track and runs up to the bridge over the expressway. Berowra station is just over the bridge.

5 BEROWRA – WARATAH BAY

Distance 5 km (return)

Time 2 hrs

Features A secluded bay surrounded by steep wooded slopes. Historic track and old boatshed site.

> *A well-defined track which descends/ascends steeply 210 m from the station to the bay by means of zig-zags.*

From Berowra station cross the narrow road bridge over the expressway. The Waratah Walking Track begins over the bridge. It is signposted. After crossing a grassy area, you enter the bush and join the original track. As you descend in a series of zig-zags, it is clear that the track has, in the past, been well made. Cut-away rock faces, sandstone embankments and stone drains are evidence of substantial work. When Windybank's Boatshed operated on Waratah Bay, this 'road' was used daily by family members as they rode their horses or walked to school or to get supplies, and by boatshed customers. As difficult as it is to believe, a Model T Ford is said to have descended

Rusting hulks

and ascended the track, being lifted by manpower around the sharp corners.

At the bottom of the slope, the track crosses Waratah Creek. Here the stream has cut deeply into the sandstone, providing shelter and moisture for many tall shrubs and trees. The wooden posts of an old bridge may still be seen.

After the creek, the remaining distance to Waratah Bay is along the gully above the stream. Bordering the track you will see *Phebalium dentatum, Asterolasia correifolia* , mint bush and pine-leaf geebung. A common small tree is grey myrtle. Further down the slope are large kanooka, coachwood and lillypilly. The remains of a water pipeline and telephone cables can be seen along this part of the eroded road.

Waratah Bay is a very pleasant spot. It was the site of one of the first buildings on Cowan Creek. Edward Windybank built a boatshed here in the early 1890s and the bay became a centre for boating activity. Boats were built, repaired and hired. At one stage, Windybank had 30 boats and 11 houseboats for hire. He ran a store with fresh daily supplies coming by packhorse from Berowra. The Windybanks lived at Waratah Bay and raised a family of seven. Water was obtained from a permanent spring which still flows behind the old boatshed. There was an orchard and an aviary, and a few cows and horses were kept. The Windybank children went to school on horseback. Edward Windybank died in 1927 and his sons ran the business until well after World War II. The building was unoccupied when destroyed by bushfire in 1972.

Today, little remains except the clearings, rock walls, some exotic plants, the hulk of one of the houseboats, water pipes and the boatshed foundations – and, of course, the track down which you have come. This was built by Windybank between 1890 and 1892.

Return the way you have come or continue around to Appletree Bay or Mt. Kuring-gai Station (Walk 4).

6 COWAN – JERUSALEM BAY – BROOKLYN

Distance 11 km (one way)

Time 4 hrs 15 mins

Features Part of The Great North Walk between Sydney and Newcastle, this track passes through a variety of vegetation types, with excellent wildflower displays in unburnt areas. Jerusalem Bay is a picturesque flooded valley. Fine views are obtained from the northern ridges over Hawkesbury River. Beginning and ending at a railway station.

Note. The walk to Jerusalem Bay and return is worth doing in its own right, especially in spring. It is 4.8 km (less than 2 hours) return.

About 5 km of this walk is along a wide service track, the rest on narrow, eroded foot tracks. The route to Jerusalem Bay is well defined and well used. North of Jerusalem Bay, in the Campbells Creek area, bushfires have made the track less clear. This walk is graded moderate to hard because of the steep ascents – the track drops 190 m from Cowan to Jerusalem Bay, climbs steeply 160 m, drops 100 m, then climbs back 110 m to a ridge along which the service trail runs. There is another descent of 130 m to Brooklyn Dam, followed by a gradual rise and a short, sharp drop down to Brooklyn.

This walk begins at Cowan station. Cross the rail line, walk north and the track begins about 50 m north of the station. It is signposted. A bridge crosses the Expressway. Soon you begin to descend, at times quite steeply. The track is eroded, but easy to follow. The bush here was unburnt by the January 1994 bushfires and thick scrub containing many types of wildflowers lines the track.

At the bottom of the slope is a creek shaded by large coachwood, forest oak and turpentine. Ferns cover the banks. The track crosses the creek and runs along the sheltered slope. From mid-August through spring, the pink flowered *Boronia fraseri* dominates this section of the track. There are also fine patches of the red and white flowered *Darwinia procera* and mauve flowered *Prostanthera linearis.*

The track drops to the shoreline of the bay where a wider track takes you around the point to the site of Rhodes Boatshed. The remains of foundations, rock walls and exotic plants from the gardens are all that are left of a centre of activity which lasted for 70 years.

George Rhodes and his family settled at Jerusalem Bay in 1895. His father had owned an orchard at Turramurra and would often row to Jerusalem Bay from Bobbin Head for the fishing. George built a boatshed and had three weekenders which were rented out; later he opened a small store for walkers and boating visitors. He formed the track up the hill to the railway and convinced the Railways Department to build a station at Cowan (opened October 1901). Although at first Rhodes had no legal tenure and paid no rent, he developed a fleet

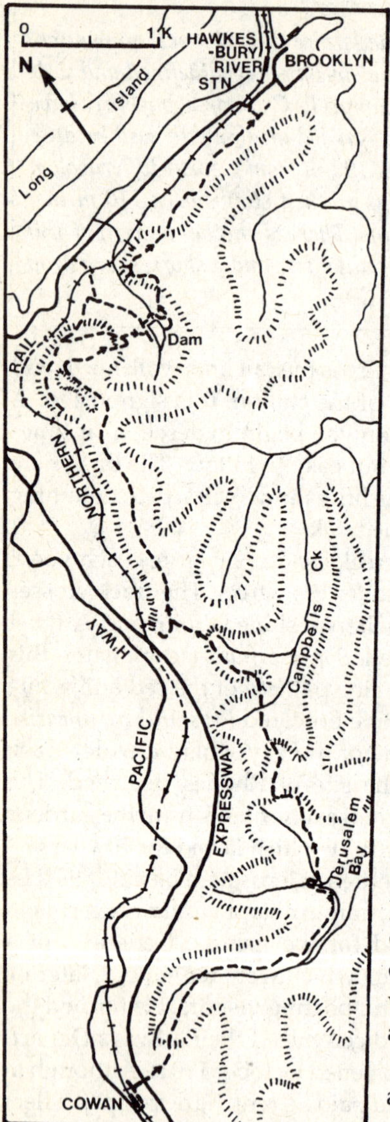

of several boats for hire and a launch. The Park Trust granted him a lease in 1901 for 5 pounds ($10) a year. Fire destroyed the Rhodes' cottage in 1908 but it was soon rebuilt. The Rhodes' children were born at Jerusalem Bay. To get to school they had to walk up to Cowan, then catch the train to Hornsby. Rhodes retained his interest in the area until after his wife died. In 1941 he sold the boatshed to Les Smith; in 1960 the business was sold to John Waters. Following a restraining order to stop the construction of a track to Jerusalem Bay, Mr Waters was evicted in 1965.

From the ruins, the track drops to the shoreline and passes around the north arm of the bay. After crossing a rocky creek, turn right to a small gully where the track begins to climb steeply. Ascend the slope, keeping the gully on your left. Near the top of the saddle, there is a prominent rock outcrop where you can sit and recover while looking out over the park and the way you have come.

Beyond the ridge, the track runs along a spur before dropping steeply to cross Campbells Creek. It then ascends and runs along a slope with a gully on your right. The January 1994 fires have affected this section of the park and care needs to be taken to keep to the track. You pass beneath a large sandstone overhang and around the upper reaches of the gully before heading westwards towards the Expressway. Ascend another slope to a wide trail which services the electricity line along a high ridge.

The walk for the next 2 km is easy along this service trail. Porto Bay can be seen to the right, the Pacific Highway and Expressway down to the left. From the northern end of the trail, you look down over Hawkesbury River, the rail bridge and across Long Island to Brisbane Water National Park. Brooklyn Dam is in the valley below. Descending from the ridge, there is a different view towards the road bridge and the river.

Follow the signposts. After a U-bend in the trail, a rough track runs down the hill to the edge of the dam. Turn left to the dam wall, then right to cross the creek below the dam. This substantial structure was built to supply water to steam engines and a pipe line runs to the railway below.

Beyond a camping site, around the edge of the water, you join up with another service trail. Turn left. It is now a little over 2 km (35 minutes) to Brooklyn along this service trail. As you approach Brooklyn, a signposted foot track runs steeply down to the main road. Brooklyn township is to your right and it is 300 m to Hawkesbury River station.

7 ELVINA TRACK

Distance 7 km (return)

Time 2 hrs 30 mins

Features Elvina Bay is a picturesque isolated settlement on the shores of Pittwater. Wonderful water views and interesting shoreline. Side track to the base of a small waterfall. Aboriginal rock engravings occur along the way.

Note. Elvina Bay can be reached by ferry from Church Point. This enables the walker to use public transport and by ascending Elvina Track to gain access to other walks off West Head Road.

Elvina Track is a wide service trail. It descends 130 m over 1 km from the plateau to the bay. The last section is quite steep. The plateau section and the bay-side section are relatively flat. The suggested (but alternative) return is by a very steep foot track which climbs 100 m directly up from Elvina Bay. Medium difficulty.

Elvina Track commences off West Head Road, 1.3 km from the Coal and Candle Drive junction. It is a well-used service trail.

Just 200 m in from the road, a side track runs off to the right to a large rock platform on which are a number of Aboriginal engravings. The turn-off is signposted. With a little searching, you should find engravings of a large three metre emu, a number of shields, a large figure with a headdress (perhaps an ancestral hero), some fish and small birds.

The far end of the rock platform has an excellent example of tessellated pavement, a geological formation characterised by a regular chequered pattern across the surface of the rock. Note also the islands of vegetation in the rock formed when a little soil and moss fill the shallow hollows. These are the habitats of a number of plants. A tiny-leaved shrub, *Baeckea brevifolia,* is most conspicuous. It often appears reddish-brown and almost dead but quickly responds to rain; it flowers irregularly, often in winter. The islands are the home of a number of inconspicuous but delicately beautiful ground orchids. Fringed midge-orchid (*Prasophyllum fimbriatum*) flowers between February and May; the greenhood *Pterostylis longifolia* flowers in May and June. Spring brings the flowering of cow orchid (Cryptostylis subulata), flying duck orchid (*Caleana major*) and bearded orchid (*Calochilus* sp.).

Return to the Elvina Track. At first it crosses the plateau but after 20 minutes you begin to descend, gently at first down a spur, then steeply down the slope. Away from the open plateau, the trees are larger and include red bloodwood, scribbly gum, smooth-barked

Tessellated pavement

angophora and old man banksia. Just before the steep final descent, there is a lookout rock with views across Lovett and Elvina Bays. A little further on, Mc Carrs Creek and Church Point are seen through the trees to your right.

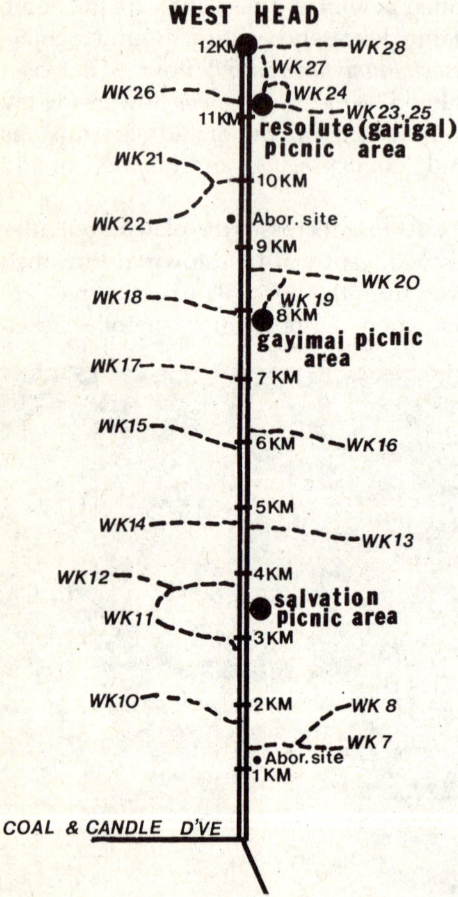

West Head

The vegetation at the base of the slope and around Elvina Bay is influenced by the richer soils derived from Narrabeen Shales which outcrop below the Hawkesbury Sandstone. Here is a forest of tall straight spotted gum, mixed with grey ironbark and forest oak. The understorey is open and contains species such as burrawang, native holly, narrow-leaf geebung, coast banksia, spine-leaved banksia, *Acacia falcata* (a wattle which flowers in June), blady grass and many smaller understorey plants.

At the base of the slope, turn left along the fire trail, passing some houses. After a few minutes you will come to a tarred path which leads down to

Elvina Public Wharf.

The fire trail continues around the bay to its northern side. Turn right at the trail junction, along the Lovett Bay side of the ridge, then over and down to another public wharf on Elvina Bay. (If you wish to catch the ferry you must signal it as it approaches). It is worth walking up to the end of Rocky Point along the trail on the middle of the ridge between Elvina and Lovett Bays. A foot track runs down through the bush to the water's edge from the side of the last house. From the rocky shoreline, you look up Lovett Bay to Salvation Creek gorge and over Pittwater to Scotland Island. Behind you, on the slope, is a spotted gum forest. Two species of mangrove – river mangrove and grey mangrove – struggle to survive among the rocks on the edge of the bay.

Return to the junction of the track which runs around Lovett Bay (10 minutes from Rocky Point). A large Swiss-style house is on the corner. Take the Lovett Bay trail along the slope behind some houses. After eight minutes you come to the base of a small waterfall beneath a sheer cliff-line. Note the presence of cabbage tree palms and, in the sheltered rocky creek below the falls, rainforest species such as coachwood, pittosporum, *Wilkiea huegeliana* and vines of *Cissus hypoglauca.*

Return to Elvina Bay and the main track. On the slope just above the track junction and the Swiss-style house is a lonely grave with a headstone engraved 'Frederick Oliver', dated 1867. The Olivers were early settlers at Elvina Bay, receiving a grant of 30 acres (12 ha) in 1865.

You may return the way you have come, but a more interesting return route is by the foot track which runs up the hill from the grave. This is strenuous, but you reach the top in about ten minutes, going around to the left of a rocky knoll. There are fine views from the top of this knoll and some good specimens of old grass trees (*Xanthorrhoea arborea*).

The track continues along the plateau. In about five minutes, you arrive at a well-defined junction. Take the right-hand branch to a creek and the top of a waterfall. This is the same waterfall which you stood under previously. It is a beautiful spot, with clear water tum-

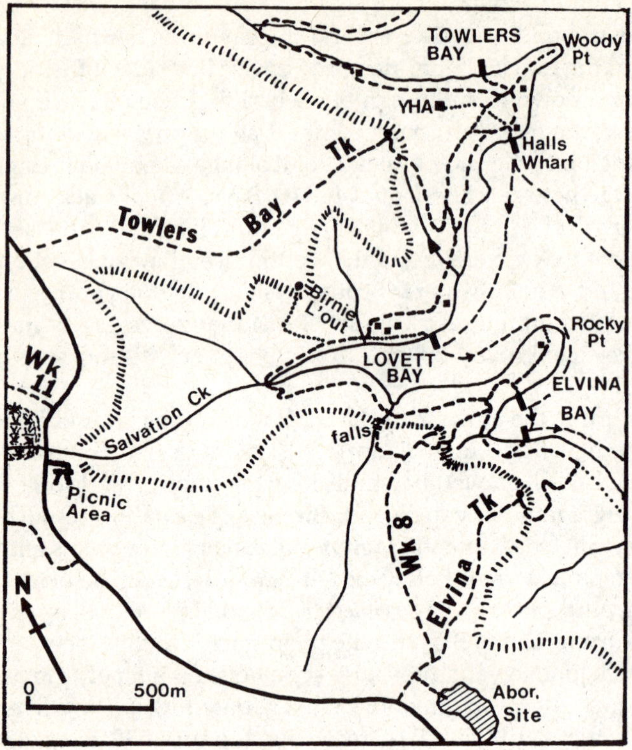

Elvina Bay–Towlers Bay

bling over rocky steps and through sculptured pot-holes. One pool is deep enough for a swim. The view across Lovett Bay from the top of the falls is one of the best in the park.

Return to the main track and follow it up the gentle slope. It is a disused and somewhat eroded service trail. In less than 15 minutes, it joins the Elvina Track. From here, it is only four minutes back to West Head Road.

8 ELVINA TRACK – WATERFALL

Distance 3 km (return)

Time 1 hour

Features An easy walk to one of the least known treasures of the park, a small unnamed creek and waterfall from which there are wonderful views over West Pittwater.

Most of the walk is along a well-defined disused service trail. There is a descent/ascent of 60 m in the 1.5 km; the gradient is gradual with no steep climbs.

This walk begins at Elvina Track, 1.3 km along West Head Road from the Coal and Candle Drive junction.

After about four minutes walk along this service trail, a smaller track runs off to the left. Take this smaller track which is an old fire access road. It was once part of a well-maintained track from Lovett Bay to Coal and Candle Creek completed in 1913. It received so little use that the Park Trust decided in 1922 that it would not be maintained. It gradually descends. In less than 15 minutes from Elvina Track, there is a track junction. Take the foot track to your left; within two or three minutes you will arrive at a creek. It is a delightful spot where water cascades over a number of rock ledges into pools and pot-holes before plunging over the cliff-line.

The view from the top of the waterfall is one of the best in the park and stretches from Lovett Bay below to the middle reaches of Pittwater.

Return the way you have come or continue for a further five minutes to the knoll above Elvina Bay (see Walk 7).

9 ELVINA BAY – TOWLERS BAY

Distance 3.5 km (one way)

Time 1 hr 30 mins

Features A pleasant coastal walk for those using public transport. Starting and ending with a ferry trip on Pittwater from Church Point, the walk runs around Lovett Bay and over the lower reaches of Salvation Creek.

A combination of fire trails and a narrow foot track. A couple of small ascents but mostly an easy gradient along the lower slopes.

From Church Point, take the ferry to Elvina Bay North Wharf. Above the wharf, turn left along an access trail which runs down along the Lovett Bay shoreline. After two or three minutes, there is a track junction, opposite a large Swiss-style house. Take the right hand fork and follow the service trail up behind some houses along a slope dominated by spotted gum, rough-barked apple, grey ironbark and forest oak.

About eight minutes from the junction, the trail ends at the rocky base of a waterfall. Note the lusher vegetation, the cabbage tree palms, coachwood and the ferns on the wet ledges.

Cross the rocks at the base of the falls. A foot track begins and runs around the slope well above the shore. In less than 20 minutes from the falls, the track drops in a series of zig-zags to Salvation Creek near the upper reaches of the bay's tidal zone. Once at the creek, walk to the left (upstream) to find the crossing and to pick up the track on the other side.

The bed of Salvation Creek is full of large mossy boulders and a rugged side creek tumbles down from the northern slope. The combination of moisture and shelter produces rainforest-type vegetation typical of such gullies in the park. The larger trees are kanooka, coachwood and grey myrtle. Lianas of water vine hang like giant snakes from the branches and two species of the vine *Smilax* can be found. Shrubs include long-leaved lomatia, small water gum (*Tristania neriifolia*), black wattle (*Callicoma serratifolia*), pepper-bush (*Tasmannia insipida*) and narrow-leaved myrtle (*Austromyrtus tenuifolia*). None of these species can be found on the drier slopes.

The track follows the northern shore of the bay. The fire of January 1994 devastated this area and destroyed a number of homes and the boatshed along Lovett Bay. At one point, about 15 minutes from the creek crossing, the track passes over a small concrete foundation and a narrow track runs off to the left. This is the route to Birnie Lookout, a beautiful spot on the edge of the escarpment. It is about 30 minutes walk to the top and very steep. If you decide not to ascend to the lookout, continue straight ahead, over a small creek and out into a clearing with some houses. Lovett Bay Wharf is down along the shore, but the main track runs directly up the slope with the creek on your left and becomes a fire trail along the slope above the houses.

As you round the first headland, you look down on a large old house which was the home of Dorothea Mackellar (1885–1968), the famous Australian poet and author of 'My Country'. A little further on, the track divides. Left and up the hill is the main fire trail. To your right, there is a service trail which degenerates into a narrow track. Both lead to Towlers Bay; the right hand route is the easier of the two.

The track drops to the headland through a beautiful spotted gum forest. Halls Wharf, your departure point, is on the Lovett Bay side of the headland and is reached by a footpath which runs to the right of a house surrounded by a fence of bamboo (see Walk 13 for a description of Towlers Bay).

10 WARATAH TRACK

Distance 9 km (return)

Time 2 hrs 20 mins

Features Ridge-top walk. Wildflower displays (depending on season and fire frequency). Views from end of track over Coal and Candle Creek.

A wide service trail, mostly easy along ridge line at 140–160 m above sea level, but passing up and over three rocky knolls, each rising 20–30 m.

The track begins off West Head Road, 1.7 km from Coal and Candle Drive junction. It is marked by the number '2' on a post.

At first the walk is across a sandy heath. It then passes over a rocky knoll (topped by Arden Trig, 206 m), along a scrubby saddle, steeply up another knoll, and back down to the ridge. There are occasional views to the left down to Akuna Bay boat facilities. To the right are the gullies of Yeomans Creek and Tully Forest. At one point there is a large rock shelf with many wildflowers growing in islands in the rock.

The plant communities are typical of those found throughout the park in similar habitats. There are many species of flowering shrubs. Some may be identified by the colour of their flowers.

- **Red flowers**: red spider flower, waratah, Christmas bell, darwinia.
- **Pink flowers**: Sydney boronia, waxflower, native rose, *Tetratheca ericifolia, Stylidium lineare.*
- **Yellow flowers**: eggs and bacon, phyllota, guinea flowers (*Hibbertia riparia, H. linearis),* sweet-scented wattle, *Pultenaea elliptica, Bossiaea scolopendria, Isopogon anethifolius, Phebalium squamulosum.*
- **Mauve flowers**: *Dampiera purpurea, Hemigenia purpurea, Scaevola ramosissima, Patersonia sericea.*
- **Blue flowers** : blue dampiera, *Lobelia dentata.*

- **Green flowers** : *Chloanthes stoechadis.*
- **Grey flowers** : grey spider flower.
- **White flowers** : woollsia, flaky-barked tea-tree, flannel flower, long-leaf coneseeds, *Leucopogon microphyllus, Epacris pulchella.*

The track passes around the northern slope of another knoll. A side track runs along a spur to the right. Take the left hand track down the slope. It is less than 15 minutes to the end of the service trail. Continue to the rocks along the escarpment. There are views north over Yeomans Bay. Further around the rocks you can see Coal and Candle Creek and the boats moored at Cottage Point. Return the way you have come.

11 SALVATION LOOP

Distance 4 km (round trip)

Time 1 hour

Features An easy walk around the swampy headwaters of Salvation Creek. This track was well known for its wonderful displays of wildflowers, but the whole area was burnt in January 1994 and it will take a few years to recover its floral diversity.

A wide service trail, generally level and with no steep grades. Some damp areas, but mostly sandy underfoot.

Begin this walk at the car park/picnic area beside Salvation Creek, 3.5 km along West Head Road from Coal and Candle Drive junction. Walk north along the road over the creek. To your left is a hanging swamp, the source of the creek. This swamp is dominated by the sedge *Gahnia,* swamp banksia (with large leathery leaves), prickly tea-tree, dog rose and coral fern.

Family on Salvation Track

The Loop Track begins 300 m beyond Salvation Creek. The service trail is wide and easy to follow. At its best, the bushland has a wonderful variety of flowering plants. There are pink, red and grey spider flowers. Sydney boronia and native rose add their splashes of pink. The white flower-heads of long-leaf coneseeds are common in spring. Sundew and dog rose thrive in the damper areas.

Fire drastically changes the species composition of the bush. Thickets of dagger hakea and heath banksia are destroyed to be replaced by fire-blackened trunks and dead stems. Those species which recover best after fire, such as dwarf apple, flaky-barked tea-tree, grass tree, honey flower and *Banksia oblongifolia,* dominate the scribbly gum woodland.

After 1.5 km, the Wallaroo Track branches off to your right. Continue around the loop. Near the end of the track, to your right, is a patch of bush left unburnt in the January 1994 fires. Compare the variety of species and number of plants in flower on this side of the track to the burnt bush on your left.

Return to your starting point by walking back along the road (eight minutes walk).

12 WALLAROO TRACK

Distance 9.2 km (return)

Time 2 hrs 30 mins

Features Changing vegetation patterns, with some good areas of healthy bushland. Wildflowers in season. Quiet areas for birdwatching. Fine views of Cowan Water from the end of the track.

Note The walk can be extended by taking the Yeomans Branch Track , an additional 3.4 km return, taking about 50 minutes.

A wide service trail, mostly easy grades along the ridge, but with some ups and downs and one short steep descent/ascent. A gradual descent of 50 m in the last 1.5 km. Medium difficulty.

This walk starts at Salvation Loop Track off West Head Road, 3.8 km from Coal and Candle Drive junction. The Wallaroo Track turn-off is 1.5 km along this loop beneath a rocky hill.

Wallaroo Track runs along the ridge between Refuge Gully and Yeomans Creek. The vegetation consists mainly of shrubs such as the yellow-flowered peas *Phyllota phylicoides* and *Dillwynia retorta*, red grevillea, grey grevillea, the white-flowered long-leaf coneseeds and heath banksia. The gullies and slopes to your left were left unburnt by the January 1994 fires and are an important refuge for small mammals and birds. Along this section of the track are some stands of the rare yellow-top ash (*Eucalyptus luehmanniana*). It is easily recognised by its mallee habit (in which a number of stems rise from a common root-stock), its yellowish young branches, silvery buds and abundant yellow flowers.

As you cross a saddle and descend a steep slope, there are distant views towards the Hawkesbury River and Refuge Bay. The track climbs back to the ridge where there is an open woodland with scattered large scribbly gum, grey gum, red bloodwood, smooth-barked angophora and stringybark. The sedge *Ptilothrix deusta* is the main ground cover and there are numerous grass trees.

The track passes along the southern slopes of Wallaroo Hill. In this shadier environment the trees are taller and include one large turpentine . The shrub understorey contains Christmas bush, broad-leaf geebung and large-leaf bush-pea. You pass back over the ridge line and arrive at the Yeomans Track junction.

It is 1.7 km to the end of Yeomans Track (a 50 minutes return walk). The route is easy, following a ridge for most of the way until it drops in an easy gradient about 40 m to a rocky platform overlooking Cowan Creek.

Wallaroo Track continues to your right for a further 1.4 km. On the way there is a good patch of heath on a rocky ridge. Some of the plants growing in this heath include darwinia, Christmas bells, red styphelia, drumsticks, native rose, tea-trees, *Pultenaea, Leucopogon, Hibbertia linearis* and the curious green-yellow tubular flowered *Chloanthes stoechadis.*

The track ends above some rocks which overlook part of Cowan Creek. This is a quiet, remote spot, excellent for wildflower photography, bird watching or just relaxing and absorbing the beauty before you. Of botanical interest is the presence of a number of small yellow bloodwood trees. This species typically grows on the edge of sandstone escarpments, but is uncommon in the Ku-ring-gai Chase. It is recognised by its soft, flaky, yellow-brown bark and its abundant flowers in October.

13 TOWLERS BAY TRACK

Distance 8.5 km (return)

Time 2 hrs 20 mins

Features Views over Pittwater; the interesting small community around the shores of Lovett Bay and Towlers Bay. Outstanding spotted gum forest at the end of the walk.

Note You can reach Towlers Bay by ferry from Church Point to Halls Wharf. The walk can then be done in reverse. You can also extend the walk by following the shoreline track around to Elvina Bay (see Walk 9).

Wide service trail for the entire route. Steep descent/ascent from the edge of the plateau to the bay, although the trail is well-graded and takes 2 km to drop 150 m.

The track begins off West Head Road, 4.7 km from Coal and Candle Drive junction. It crosses an undulating fire-affected plateau and after 2 km (30 minutes walk) comes to the edge of the escarpment. From a rocky lookout there are fine views over Towlers Bay, Scotland Island and much of Pittwater.

The trail descends. A small gully is on your right. Near the bottom of the hill, this creek has been dammed to form a small reservoir for fire fighting. The trail makes a U-turn and follows the contour of the slope parallel to but well up from the shore. A side trail runs to the right down to Lovett Bay. Keep straight ahead until the next track junction, where you bear right.

The trail descends through beautiful, tall, straight spotted gums to the headland known as Woody Point. Halls Wharf is reached by a footpath which runs to the right of a house surrounded by a fence of bamboo. The wharf is a good spot to relax, view the bay and , perhaps, to have lunch. A regular ferry calls at the wharf and, if time permits, you could take it to Elvina Bay or Lovett Bay and walk back following Walk 9.

A service trail runs over the ridge to a small wharf on the Towlers Bay side. Another runs up the slope to a youth hostel and a third trail takes you around the slope to the upper reaches of Towlers Bay.

The spotted gum forest on Woody Point is one of the finest examples of such a plant community in the Sydney area. It is the result of outcropping shales which produce richer soils. The spotted gums are easily recognised by their straight trunks with a mottled appearance. Grey ironbark and forest oak also like these richer soils, while burrawang, hop goodenia and native holly are all common.

Return the way you have come.

14 WILLUNGA TRIG

Distance 1.8 km (return)

Time 30 mins

Features A short, easy walk to the highest point in the park (246 m). Views in all directions from the rocky hilltop.

A wide service trail which ascends 70 m from West Head Road, gradually at first, but steeply at the last section.

The track leaves West Head Road on a saddle, 4.7 km from Coal and Candle Drive junction. It is marked by a number '6' on a post.

Near the start of the track are patches of bush unburnt in the January 1994 fires. Here you will find plants typical of the dry ridge tops, such as red grevillea, flaky-barked tea-tree, darwinia, banksias (*B. oblongifolia* and *B. ericifolia*) and dwarf apple.

The track ascends to a saddle through fire-affected woodland, then climbs steeply to the rocky top. From here you can look east to Pittwater, west to Cowan Creek and north and south over a large expanse of the park.

15 TOPHAM TRACK

Distance 3.6 km (return)

Time 1 hour

Features An easy walk through open woodland to a viewing point over Cowan Water, Refuge Bay and America Bay.

A wide service trail with a few ups and downs but no steep climbs. An easy descent of 40 m in 1.8 km.

The track begins off West Head Road, 5.9 km from Coal and Candle

Drive junction. It is marked by the number '7' on a post.

The first part of the track skirts Topham Hill. The vegetation of the slopes is influenced by the thin sandy soils, the northern exposure to the sun, run-off from the hillside and, above all, by bushfires. Tea-trees, dwarf apple and *Banksia oblongifolia* all have large underground root-stocks which allow the plants to survive fire; they send up new shoots within weeks of being burnt. Black she-oak, needle-bush, silver banksia and heath banksia are killed by fire and must grow again from seed. Scribbly gums renew themselves after fire by epicormic growth in which shoots sprout along the old wood of the trunk.

From higher points there are views over the gullies and distant bays. The bushland of the ridge and slopes consists of fire-damaged woodland dominated by small scribbly gums and red bloodwoods. The shrub understorey includes tea-trees, banksias, grass trees, prickly moses, broad-leaf geebung, broad-leaf coneseeds, wax flower and an occasional soft boronia.

In less than half an hour, you arrive at the end of the track. From rocks on the edge of the escarpment, you look over the entrance to Refuge Bay (to the left) and America Bay (to the right).

16 BAIRNE TRACK – TOWLERS BAY LOOKOUT – SOLDIERS POINT

Distance 10.8 km (return)

Time 2 hrs 20 mins

Features An easy walk along ridge tops to two of the great lookout points in the park. Bairne Track leads to Towlers Bay Lookout (3.7 km each way) and a spur track runs to Soldiers Point (an additional 1.7 km each way).
Walkers who want something more strenuous can extend this walk by taking one of three narrow, steep and rough tracks down to the shore: to Portuguese Beach, to Towlers

Bay foreshore from the lookout or to Bonnie Doon Wharf on the southern side of The Basin. By catching the ferry from Bonnie Doon across to The Basin Wharf, you can return to West Head Road by The Basin Track (Walk 20).

Note. Access to these tracks can be gained by ferry from Palm Beach (Pittwater) to Bonnie Doon Wharf.

Service trails along ridge. Some minor ups and downs, with a gradual drop of about 70 m at the end of Soldiers Point Track, but no strenuous ascents.

Bairne Track commences off West Head Road, 6.1 km from Coal and Candle Drive junction. It is signposted and has the number '8' on a post. The wide service trail runs through heath and scribbly gum woodland along a fairly level ridge. It skirts the knoll on which Bairne Trig (183 m) stands. In less than 30 minutes walk from the road, a foot track runs off to the left. This is the track to Bonnie Doon Wharf. Continue along the service trail, down an incline to the junction of Portuguese Track and Soldiers Point Track.

Keep to your right. Towlers Bay Lookout is 13–14 minutes away. Here on a rock perched 110 m above the bay, and right on the cliff edge, you look out over Pittwater, Scotland Island, Newport Peninsula, Towlers Bay, McCarrs Creek and further south. It is a spectacular view and you will want to remain for a while absorbing the grandeur. It is possible to descend by a narrow track to the foreshore of Towlers Bay. The track begins about 250 m back along the main track.

The track from the foreshore to the lookout was completed in June 1910 and extended to The Basin in 1912. In 1900 there was a floating powder hulk in Towlers Bay and an Explosives Reserve on shore. A cottage was built for the employees of the Explosives Department. This was eventually handed over to the Park Trust. Between 1930 and 1940, a park ranger, Len Lovegrove, lived in a cottage below the lookout. To get to work each day, he walked to and from The Basin and the track became known as Lovegrove's Track. The Trust's cottage was used as a youth hostel until 1967. Subsequently, it was destroyed by fire.

Return to the Soldiers Point junction. Turn right. After 200 m, the Portuguese Beach Track runs off to your right. Take the left hand service trail. This travels for 1.5 km along a ridge, gradually dropping to another fine lookout which looks directly up into The Basin, over Coasters Retreat to Currawong Beach and north to Lion Island. Return the way you have come.

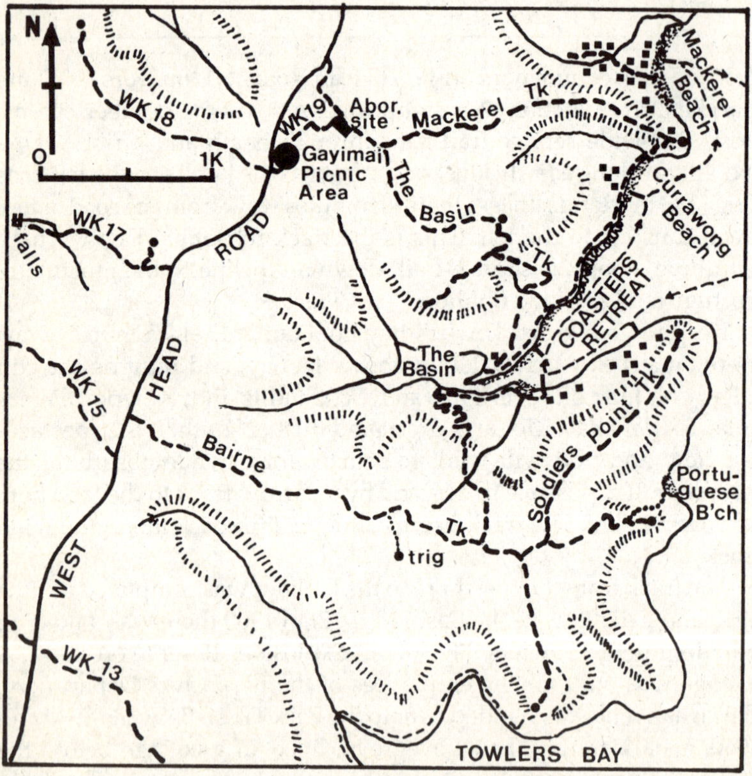

Central Lambert Peninsula

Portuguese Beach. Only fit walkers with sturdy boots should attempt this walk as the route down the slope (80 m vertical descent) is a scramble down a steep, rough and poorly defined track, made more obscure by fires. The track at first is a service trail running off Soldiers Point Track. It follows a spur for 700 m, dropping 60 m to the edge of the escarpment. There are views of Pittwater from the rocks at the end of the trail. The narrow foot track begins to your left near the end of the service trail. Give yourself 20 minutes to walk to and from the end of the service trail, and 50–60 minutes to Portuguese Beach (return).

The Basin (Bonnie Doon Wharf). This narrow foot track descends 150 m in a distance of 2 km from the Bairne Track to the water's edge. The first part is down over a creek and around a swampy slope. The final descent is via a series of zig-zags down a very steep slope, then along the southern shore of The Basin, past the shark net and around the rocks to Bonnie Doon . Regular ferries leave this wharf for the other side of The Basin, Currawong, Mackerel Beach and Palm Beach. It takes about 30 minutes to descend from the plateau and 40 minutes to ascend.

Walkers who catch the ferry to Bonnie Doon and walk uphill may have a little difficulty finding the track. Turn right at the wharf, cross a creek and the small Corner Beach, then follow the shoreline around past the shark net. The track goes along the shoreline for about 150 m beyond the net, then begins to ascend.

17 AMERICA BAY TRACK

Distance 2 km (return)

Time 1 hour

Features An easy to moderate walk to a viewing point and small waterfall on the edge of the escarpment overlooking picturesque America Bay. An Aboriginal rock engraving site can be visited along the way.

The bushland along this route was badly burnt in the

January 1994 fires and will take some years to recover its former glory.

A narrow track with a descent of about 80 m. The lower sections of the track are rough and badly eroded.

The track begins off West Head Road, 6.9 km from Coal and Candle Drive junction. It is marked by the number '9' on a post.

About 90 m in from the road, a side track runs downhill for 70 m to a rock flat where there are a number of Aboriginal engravings, including a goanna, a ray, a large shark and a female figure. A natural water soak runs across part of the rock, filling a pot-hole around which are some faint axe-grinding grooves.

Return to the main track which gradually descends through a fire-affected woodland of scribbly gum, Sydney peppermint, red bloodwood, dwarf apple, old man banksia and grass trees. In damp areas, sedges and pink swamp heath dominate and small red leaf rosettes of the insectivorous sundew are common beside the track.

At the bottom of the slope is a small creek with sparkling clear water. The track follows this creek, then crosses it. You soon arrive at the lookout on the top of a large rock shelf where a waterfall tumbles into the gully below. Above the falls are small cascades, rock holes and large pools. Water gum and two species of bottlebrush grow in the wet protected environment below the cascades.

There is a fine view across the narrow America Bay to Challenger Head. The name 'America Bay' was first mentioned on a survey completed in 1832 by Surveyor Larmer. The headland is named after HMS *Challenger* which visited the area between 1866 and 1870.

It is possible to get down to the water's edge, although the route is steep and eroded. There is no marked track, but you can descend by a rocky watercourse which begins about 150 m around the cliff to the left of the waterfall.

18 CHALLENGER TRACK

Distance 3 km (return)

Time 1 hour

Features An easy walk to a viewing point overlooking Cowan Water and the Hawkesbury River. Wildflowers in season. The bushland along this route was badly burnt in January 1994 and will take some years to recover its floral diversity.

A wide service trail, mostly flat and along the ridge line. No steep climbs.

The track leaves West Head Road opposite Gayimai Picnic Area, 7.9 km from Coal and Candle Drive junction. The ridge top has sandy soils and many rocky outcrops so the vegetation is a scrubby heath with scattered small scribbly gums and red bloodwoods. The presence of wildflowers depends on the frequency of fire. Banksias, tea-trees, honey flower and dwarf apple are common. The grass tree (*Xanthorrhoea media*) flowers prolifically in autumn and winter after summer fires.

The main trail follows the ridge and bears north around a prominent knoll. Here the trees are taller. Large examples of red bloodwood, smooth-barked angophora, turpentine and mahogany are present. The trail ends at a rocky lookout on the edge of the escarpment. From here you can look up Hawkesbury River to Brooklyn Railway Bridge and across Cowan Water and Broken Bay. Juno Point is prominent to the north.

19 GAYIMAI – ENGRAVINGS

Distance 1 km (return)

Time 25 mins

Features Easy access from Gayimai Picnic Area to one of the best Aboriginal rock engraving sites in the park.

A narrow, eroded foot track, with an easy ascent of about 30 m to the Basin Track and the engravings site.

Gayimai Picnic Area is off West Head Road, opposite Challenger Track, 7.9 km from Coal and Candle Drive junction.

The track begins at the northern end of the picnic area and makes its way through fire-affected scrub and scribbly gum woodland dominated by banksias, needle-bush, tea-tree and dwarf apple. About 50 m from the start is a Y-junction; bear left. Soon you ascend the slope to the plateau where you will find a wooden walkway to the rock platform on which the engravings are located.

The engravings are in three groups. Signs have been erected to help you interpret what you are looking at and to explain the engraving process. In Group A, there are a number of figures, including a male with a headdress holding a fish and a boomerang , and a female with a coolamon (a basin-shaped wooden dish made and used by the Aborigines). Note the belts on the men and the carved footprints across the rock. Group B consists of a large fish and the outline of a man with a fish inside his body. Group C has a number of fish, a row of hopping wallabies, male and female figures with outstretched arms, a boomerang and other less distinct figures.

The Aboriginal site is beside The Basin Track so you can continue your walk, or return the way you have come.

20 THE BASIN TRACK – CURRAWONG – MACKEREL TRACK

Distance 7.3 km (return)

Time 2 hrs 20 mins

Features The Basin picnic area has a beach and netted swimming area. Currawong is a pleasant beachfront property with rental cottages. Great views of Pittwater, especially from the ridge above Currawong Beach. Aboriginal engravings along the track. Wildflowers in season.

You may wish to walk only to The Basin (5.6 km return) or only along Mackerel Track to the viewing point on top of the escarpment (5.2 km return). The latter track avoids the steep descent to the shoreline.

Note. The Basin and Currawong can be reached by ferry from Palm Beach (Pittwater). This enables the walker to use public transport to complete this walk.

A wide service trail to The Basin, then around the rocks of the shoreline to Currawong Beach (best done at low tide). A steep ascent of 50 m via a narrow track to the ridge top and escarpment between Currawong and Mackerel Beach. Return by track and Mackerel Service Trail. The descent/ascent between West Head Road and the shoreline is 150 m and involves quite a steep drop from the plateau.

The Basin Track leaves West Head Road 8.5 km from Coal and Candle Drive junction. Within five minutes, you reach a fine set of Aboriginal rock engravings. Access is by a wooden walkway. The NPWS has erected explanatory signs (see Walk 19).

Four minutes further on, a service trail runs off to your left. This is Mackerel Trail, your return route. The Basin Track continues over the fire-affected plateau. Just before you begin to descend steeply, there is a view across Broken Bay and Barrenjoey Peninsula.

Pelicans – The Basin

Fire has badly affected the tall forest of smooth-barked apple, red bloodwood and Sydney peppermint on the slopes. After you pass the Euro Track turn-off on your right, the slope becomes steeper and the vegetation changes. Here on the richer soils and on the more sheltered slope are tall grey ironbark, mahogany, forest oak and turpentine. The understorey is open; grass trees are common. At one spot, the track crosses a small watercourse where there is rainforest-like growth of lillypilly, coachwood and Christmas bush. A small Port Jackson fig clings to the damp rocks. Ferns are abundant and there is even an epiphytic orchid *Liparis reflexa.*

The Basin consists of a sandy beach and an inner lagoon surrounded by 100 metre wooded hills connected to Pittwater by a narrow channel. Steam boats once passed through this channel to a wharf in the lagoon, but today it is quite shallow and is spanned by a shark-proof net. There are large grassed camping and picnic areas sheltered by Norfolk Island Pines. It is an attractive area with yachts and cruise boats always tied up in the inlet known as Coasters Retreat. Historic Beechwood Cottage, built in 1882 by Frederick Jackson, has been preserved. Facilities include a modern toilet block, drinking water, telephone, barbecue areas and a park office. Intending campers should first ring the ranger-caretaker on 02 974 4036.

The Basin was dedicated for camping and public recreation in 1879, fifteen years before the formation of the Ku-ring-gai Chase. Near the beach, you will see a small monument erected in 1922 by yachtsmen to the memory of Mrs Peggy Morris who lived at this spot for over 50 years. Her one-roomed residence was on land purchased in 1914 for addition to the park but the Trustees came to an arrangement to allow Mrs Morris to remain for the payment of five shillings a week. Her yachting friends helped her build a new cottage on the site of the old one where she remained until her death in June 1921, aged 84 years.

There is a wharf at The Basin. You may find it easier to catch the regular ferry to Currawong Beach, especially if it is high tide. The walk along the beach and around the rocks takes about 20 minutes at low tide.

Currawong has a beach and a grassed valley with a small six-hole golf course. Coral trees attract lorikeets and honeyeaters; glossy cockatoos sometimes feed on the casuarina nuts. A number of cabins sit on the lower slopes (phone 02 974 4141 for details). Currawong, originally known as Little Mackerel Beach, is not part of the park as it was an early land grant. A block of 40 acres (16 ha) was selected for a farm by John Clarke between 1823 and 1832. No developments were made and in 1835 the block was granted to Martin Burke, the so-called 'Father of Pittwater'.

At the northern end of the beach, a rough track begins, ascending the slope to the ridge. A large rock juts from the edge of the escarpment, providing wonderful views of Pittwater.

The return is along a ridge track which climbs steadily. There are views down over the village of Great Mackerel Beach. The foot track becomes Mackerel Service Trail higher up and crosses the plateau back to your starting point.

21 FLINT AND STEEL TRACK

Distance 2.6 km (return)

Time 1 hour

Features Pleasant, picturesque beach backed by steep wooded slopes. Panoramic views over Broken Bay.

A narrow, rough foot track. A very steep descent, dropping 140 m from the car park to the beach.

The track begins at a car park to the left of West Head Road, a little over 1 km from Resolute Picnic Area. It is marked by the number '13' on a post. You descend immediately. Keep left below the first rock outcrop and walk around and down the slope keeping the gully on your right. After seven or eight minutes, there is a track junction. Take the right hand track to Flint and Steel (White Horse) Beach.

The track runs along the slope of a rocky knoll. Forest oak and rough-barked apple are common. Cabbage tree palms grow in the richer soils of the gully. Native pigeons are sometimes seen feeding in the palms. Caves, weathered overhangs and boulders are conspicuous along the rock outcrop on your left.

After skirting the slope, the track drops steeply. As you come down to Flint and Steel Point, a fine panorama greets you. North is Lion Island and Middle Head. Across Broken Bay is the beach and the village of Patonga backed by the bushland of Brisbane Water National Park. The point itself has low vegetation and scattered coast banksia, well adapted to the winds which blow in from the sea.

Drop down to the beach, a beautiful stretch of yellow sand. Explore the rock shelf at the western end of the beach and note the remarkable weathering of nearby sandstone rocks.

Return the way you have come.

22 HUNGRY BEACH TRACK

Distance 3.4 km (return)

Time 2 hours

Features A combination of bush, foreshores and extensive views, ending at a small and isolated beach.

A narrow, rough foot track, washed out in places. Fire has made the lower section of the track a little obscure. A steep descent/ascent of 140 m from the road to the bay. Some coastal rock hopping near the end.

The track begins at a car park to the left of West Head Road, a little over 1 km from Resolute Picnic Area. It is marked by the number '13' on a post.

The track descends immediately from the car park. Keep left below the first rock outcrop and walk around and down the slope keeping the gully on your right. After seven or eight minutes, you come to a track junction – left to Hungry Beach, right to Flint and Steel Beach (White Horse Beach). Note the large forest oaks on either side of the track and the cabbage tree palms in the gully and on sheltered slopes.

The Hungry Beach Track descends the slope beneath forest oak and rough-barked apple. Large grass trees with fire-blackened trunks are common. On your left is a rocky outcrop with a sandstone cliff-line. The whistling kite, a large brown eagle-like bird, is often seen overhead as it nests in the area in winter. At one point, you pass a small rock well, formed by damming a natural spring in a cave.

Near the shoreline, the track goes around the ruins of a shack, through exotic vegetation which has escaped from the old garden, and down to a small patch of sand. You could stop here or continue to Hungry Beach, about 20 minutes further on. An overgrown and eroded track ascends from the sand and runs parallel to the shoreline. As it nears Hungry Beach, the track drops to the shore and you need to negotiate your way around the rocks for the final 200 m.

Hungry Beach is a pretty spot, a popular call for weekend boaters.

The concrete foundations of a wharf sit at the northern end and a freshwater stream flows down the hillside and across the sands. The beach once had some old naval buildings, but the coral trees and some rubble are all that remain.

Retrace your steps to your starting point.

23 NORTH MACKEREL TRACK

Distance 3 km (return)

Time 1 hour

Features An easy walk along a wide track. Excellent wildflower area. Panoramic views from the end of the track over Pittwater. Aboriginal rock engravings occur beside the track.

A wide service trail with a gradual descent of about 50 m from start to finish. No steep gradients.

The walk begins at Resolute (Garigal) Picnic Area, 800 m before West Head Lookout. The start of the track is above and to the right of the toilet block and is clearly signposted as 'Resolute Track'. The first section of the walk follows the same route as the Headland Loop, with the North Mackerel Track branching off to the right after 600 m.

The track passes through heath and shrubland rich in flowering plants. The taller shrubs are heath banksia, tea-tree, black she-oak, native broom and dwarf apple. Smaller flowering shrubs include red grevillea, grey grevillea, phyllota, sweet-scented wattle and mauve-flowered *Hemigenia purpurea.* Native rose, blue dampiera and purple iris all occur in the understorey.

After five minutes walk, you arrive at a small set of Aboriginal rock engravings. The site is marked by a NPWS sign. On an expanse of flat

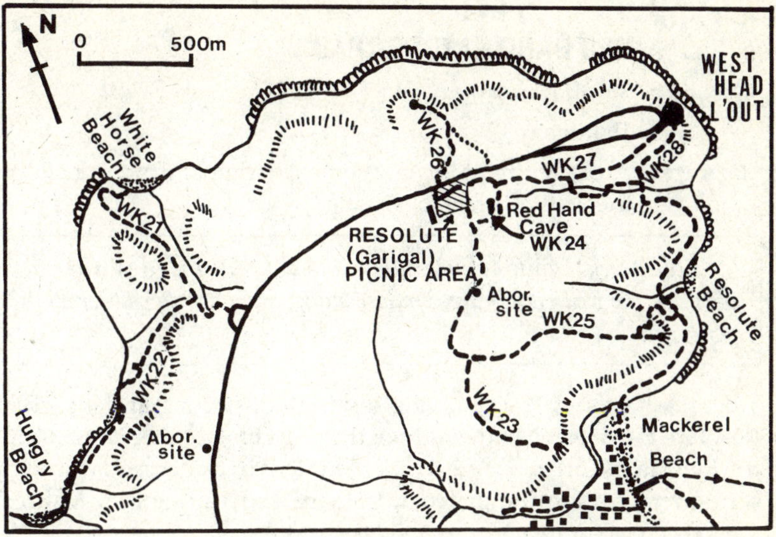

West Head

sandstone, you will find a male figure, an object resembling a spherical-headed club, two eel-like figures, a fish and part of a fin of a shark or whale. There is also a good view over Pittwater from the site.

An additional five minutes walk brings you to the North Mackerel Service Trail turn-off. After ten minutes of gradual descent, this trail ends in a turning circle. Scramble to the end of the rocks to your left for a spectacular view down to Great Mackerel Beach and across Pittwater to Palm Beach. Barrenjoey Head can just be seen to the far left. You will want to spend some time here before retracing your steps.

24 RED HAND NATURE TRAIL

Distance 300 m (return)

Time 15 mins

Features An easy circuit to view some Aboriginal hand stencils on the wall of a rock overhang.

A narrow track beginning and ending at Resolute Picnic Area, dropping a little to the Red Hand Cave and ascending via a few steps back to the picnic area.

Take the Resolute Track, a wide service trail starting at the back of Resolute Picnic Area to the right of the toilet block. After about 60 m, the Red Hand Circuit leaves to your left. This track descends through scrubby vegetation dominated by flaky-barked tea-tree, pink tea-tree, dagger hakea, prickly moses and heath banksia.

After only two or three minutes from the turn-off, you come to a rock outcrop. Look closely and you will see some outlines of red hands on the sandstone walls. These 'paintings' were made by blowing liquid red pigment over the outstretched hand as it was held against the rock. It is suggested that this cave art site might be 2000 years old. There are also some examples of more modern graffiti which detract from the significance of the site. Please do not touch the rock surface or disturb the area in any way.

The rock outcrop is on the sheltered side of the slope. Note the taller vegetation and the presence of such species as blueberry ash (a small tree with white flowers and blue berries), turpentine (a tree with fibrous bark which prefers good moist soils), the tall yellow-flowered pea *Pultenaea flexilis* and the climber sweet sarsaparilla.

Pass the track to West Head and ascend the steps back to the picnic ground.

25 RESOLUTE TRACK – MACKEREL BEACH – HEADLAND LOOP

Distance 7.3 km (round trip)

Time 3 hours

Features A photographer's dream – scenic views over Pittwater, the small picturesque Resolute Beach, the interesting isolated settlement of Great Mackerel Beach, a small set of Aboriginal rock engravings and a variety of plant communities from dry heath to rainforest gullies.
The walk can be shortened by omitting the side track to Mackerel Beach and restricting your route to Resolute Track and Headland Loop (4 km, 1 hr 40 mins return).

Note. Mackerel Beach can be reached by ferry from Palm Beach (Pittwater). This enables the walker to use public transport and by following the track in reverse to gain access to the West Head area.

The first section (1.4 km) of Resolute Track is along a service trail. The remaining distance is along narrow foot tracks. The track to Resolute Beach and the Headland Loop are well defined and maintained and have steps constructed to help descents/ascents. About 1.8 km of the route is on narrow, poorly maintained tracks. There are some steep sections as the drop from the Picnic Area to Pittwater foreshore is 170 m. Medium difficulty.

Resolute Track begins at Resolute (Garigal) Picnic Area, 800 m before West Head Lookout. Follow the wide service trail down past the Aboriginal engravings site to the North Mackerel Service Trail (as described in Walk 23). Continue along the main trail past the North Mackerel turn-off.

The trail drops gradually down a spur through fire-affected scribbly gum woodland. There is a rich understorey of flowering plants,

Mackerel Wharf

including woollsia, pinnate boronia, red spider flower, guinea flower, eggs and bacon, phyllota and darwinia. Grass trees (*Xanthorrhoea media)* are plentiful and conspicuous when in flower.

The fire trail ends on the edge of the escarpment and steps lead down to your left. Before you descend, go straight ahead to a rocky lookout from where there is a great view over Pittwater, Barrenjoey and north to Bouddi. The lookout is surrounded by large smooth-barked apple, red bloodwood and mahogany (*Eucalyptus umbra*). There are also fine examples of the larger of the two species of grass trees in the park, *X. arborea.*

The steps take you down a very steep slope, sheltered by forest oak and large smooth-barked apple. Ferns dominate the understorey. You pass a large sandstone cave which is an old Aboriginal occupation site. From beneath its shelter, there is a view directly across to Barrenjoey. A NPWS sign gives some details of the site. Near the bottom of the slope, you join the foreshore track. To the right is

Mackerel Beach, to the left is Resolute Beach, West Head Lookout and Resolute Picnic Area via the Headland Loop.

Mackerel Beach Track is narrow and a little rough. It runs around the slope above the shoreline and there are good views through the trees. After about 12 minutes from the junction, you drop to the northern end of Mackerel Beach where you must cross a small stream. At high tide you will need to take off your shoes and socks and wade across. Walk along the beach to the wharf.

Great Mackerel Beach is a pleasant holiday community with no vehicular access. Quite a number of houses are located on the flats behind the beach and up the valley. There are no shops in the village but there is a public telephone near the wharf. The area was never part of the national park. As early as 1835, 60 acres (24 ha) of land behind the beach were granted to a Martin Burke who also had an adjoining 40 acres (16 ha) south to Currawong.

It is worth having a look around the village. Take the grassed 'road' from behind the Submarine Cable sign on the beach, cross the bridge over the creek and continue up beside the creek to the last house, then return to the wharf. There is no way through from the southern end of the beach as it is all private property.

Return to the Resolute Track. Within 100 m, the track divides. The main Headland Loop goes to the left down over a rocky creek. To the right, down steep steps is a picturesque arc of sand known as Resolute Beach. The rocky creek plunges down the slope beside the track. Because of the shelter and moisture, it supports a number of rainforest species such as coachwood, grey myrtle, lillypilly, cabbage tree palm, kanooka, Port Jackson fig, brush muttonwood and the climber *Cissus antarctica.* Closer to the beach are tall rough-barked apple; coast banksia encircle the beach. This beach is believed to be the one on which Captain Phillip and his party camped during the stormy night of 5 March 1788 while exploring Broken Bay and Pittwater.

Return up the steps to the main track, cross the creek and follow the slope around through forest oak, grass trees, woolly star-hair and hop goodenia to another small creek. A set of steps takes you to the top of a rock shelf. You then descend and cross the creek. Over the

Lookout on ridge above Mackerel Beach

creek, the track divides. Right is West Head Lookout, left up the gully is Resolute (Garigal) Picnic Area. Take the route up the gully.

This track is narrow and not so well defined as the Headland Loop. You re-cross the creek at the top of a small waterfall in front of a small rock dam. Ascend, keeping the creek on your right. Coachwoods dominate the creek; the slope has a number of flowering shrubs, including *Boronia fraseri.* After 15–20 minutes walk from the Headland Track junction, you again cross the creek and join the main track from West Head. Turn left. Your starting point is 500 m up this track.

26 KOOLEWONG TRACK

Distance 800 m (return)

Time 30 minutes

Features A short walk to a wonderful viewing spot overlooking Broken Bay and Lion Island.

A narrow walking track which makes a short but sharp descent of 70 m to a rock platform on the edge of the escarpment. The return is a little strenuous.

The track begins opposite the Resolute (Garigal) Picnic Area. It is marked by the number '17' on a post. It descends through tall iron-bark, turpentine and forest oak. These slopes and the understorey were burnt in the January 1994 fires.

Nearer the lookout, the vegetation has not been recently burnt and there are large old flaky-barked tea-trees, Sydney boronia, wax-flower, grass trees, red bloodwood and smooth-barked angophora with huge trunks and spreading canopies, and a stand of native cypress (*Callitris muelleri*).

The viewing point is a magnificent spot and you will want to spend some time there. The roots of an angophora twist over the rocks and down into the crevices. Wildflowers surround the platform in spring. The flower spikes of coast banksia are conspicuous just below the edge of the escarpment. Beyond the mouth of the Hawkesbury River are the wooded ridges of Brisbane Water National Park. Lion Island squats like a sentinel at the entrance to the bay.

27 RESOLUTE PICNIC AREA – WEST HEAD LOOKOUT

Distance 1.8 km (return)

Time 1 hour

Features A stroll through sheltered eucalypt forest to West Head Lookout. Good for plant identification and bird watching. An easy downhill walk if you can arrange a vehicle to pick you up at the lookout.

A wide, well-defined track with steps constructed where necessary. The gradual descent of 70 m is most noticeable on the way back.

This walk commences at Resolute (Garigal) Picnic Area off West Head Road, just 800 m before West Head Lookout. The broad track begins between the road and the toilet block. Descend the steps. After only a few m, the track to the Red Hand Cave branches to the right. It is only 60 m along to this site.

The West Head Track follows the contours of a south-facing slope and keeps below the ridge. The slope is sheltered from the winds and from the direct rays of the sun for most of the time. The vegetation consists of an open forest of Sydney peppermint, smooth-barked apple and smaller turpentine, together with a tall shrub understorey of blueberry ash, Christmas bush, flaky-barked tea-tree, narrow-leaf geebung and the pea flower *Pultenaea flexilis.* Smaller plants include *Boronia fraseri, Pultenaea ferruginea,* spine-leaved banksia, prickly moses and crinkle bush.

The mixture of tall eucalypts, rocks, scrub and nearby creek makes this a good area for bird watching. Brown thornbill, striated thornbill and rainbow lorikeet can be seen in the canopy. Noisy miner, spinebill, yellow robin, grey fantail, little wattlebird, golden whistler and white-cheeked honeyeater flit through the understorey. The sharp call of the whipbird is frequently heard from the gully.

The track descends gradually to emerge from the bush at West Head Picnic Area and lookout. The view across Pittwater, Barrenjoey and Broken Bay is unforgettable and is one of the main tourist attractions of the park.

In 1940, during World War II, the military located two large guns in concrete emplacements near the shoreline of West Head. There were also observation posts, search light positions and a train line connecting the gun sites to a military camp on the headland. The cutting for the train line is still visible. After the war, the National Fitness Council purchased the buildings for a camp. In 1951, this West Head land came under the control of the Park Trust. The camp was closed in 1964, about the same time as the West Head Road was opened to allow the public easy access to the lookout.

Return by the same track. An alternative, but longer and more strenuous route, is to take the Resolute Beach Track (Walk 28). At the bottom of the first gully, another track leads up the gully (Walk 25) back to the West Head Track. This alternative adds about 900 m and 20 minutes to your walk.

28 WEST HEAD LOOKOUT – RESOLUTE BEACH

Distance 2.4 km (return)

Time 1 hour

Features Views of Pittwater, rainforest gully vegetation and the picturesque Resolute Beach.

Foot track, rough in parts, but mostly clearly defined. Wooden steps down to Resolute Beach. A descent of 100 m in two steep stages.

From West Head Lookout at the end of West Head Road, take the track downhill to the right of the toilet block. Minor tracks run to the edge of the escarpment but the main route bears to the right and passes down a cleft in the cliff-line. It then runs down the sheltered slope to a small gully where there is a track junction. Take the left hand track; it crosses the creek, goes up over a rock, then descends beneath the falls by wooden steps.

The track now keeps roughly to the same level as it rounds the headland. Forest oak is common. The understorey consists mainly of *Astrotricha floccosa*, an erect shrub with woolly stems, large leaves and terminal clusters of woolly white flowers borne between October and January.

Soon after crossing another shaded, rocky creek, the track to Resolute Beach branches to the left and descends via a set of well-made steps (see Walk 25 for a description of the creek and beach).

29 CENTRE TRACK

Distance 3.2 km (one way)

Time 1 hour

Features An easy walk through an area rich in wildflowers. Views of the park from the high ridge. It is best to arrange to have a vehicle meet you at the end of the walk near the junction of Coal and Candle Drive and West Head Road.

A wide service trail along a ridge. Mostly flat except for a 20 m ascent around a knoll and a short sharp descent at the end.

Centre Track begins to the right off Coal and Candle Drive just 30 m beyond the park entrance toll booth, near the corner of McCarrs Creek Road. There is roadside parking before the toll booth.

The track passes along a ridge which is 160–180 m above sea level, with views to the right over McCarrs Creek catchment and to the left over the valley of Coal and Candle Creek. This route provided the original access to West Head as it avoided the descent to McCarrs Creek. Parts of the rock walls of the original road can still be seen in places.

There are many small flowering shrubs in the sandy exposed heath. Towards the end of July, the area comes alive with colour.

There is red spider flower and darwinia, pink Sydney boronia and waxflower, white *Epacris, Leucopogon, Zieria laevigata, Conospermum ericifolium* and flannel flower, yellow *Hibbertia, Phyllota, Pultenaea, Acacia* and *Phebalium,* grey spider flower, blue *Dampiera* and purple *Patersonia.* Scattered scribbly gums, grass trees and dwarf apples are also common.

You pass over or near a number of rock platforms which not only provide viewing points but also protect plants in small islands of soil. Sedges dominate damper areas. In wet peaty soils, you will find the tiny mauve-flowered bladderwort (*Utricularia*), an insectivorous plant using tiny underground bladders to trap microscopic organisms. Two other insectivorous plants are also common – the sundews *Drosera spathulata* and *Drosera peltata,* which use sticky leaves to trap small insects.

You ascend a small knoll near the end of the track . From the rocky escarpment to the left, there is a good view over Coal and Candle Creek. Here, too, on the tessellated sandstone are some very faint Aboriginal 'footprint' engravings.

Return the way you have come, or descend to Coal and Candle Drive. West Head Road is just 300 m up the hill to your right.

Additional Walks for the More Adventurous

Each walk has public transport to its start and at its end.

- 30. North Turramurra – Sphinx Track – Bobbin Head – Appletree Bay – Berowra. **Distance :** 14.5 km

- 31. Mt Kuring-gai Station – Appletree Bay – Bobbin Head – Gibberagong Track – Grosvenor Track – Grosvenor Street – Wahroonga Station. **Distance :** 14 km

- 32. Church Point – ferry – Elvina Bay (North) Wharf – Lovett Bay – Towlers Bay Track – West Head Road – Elvina Track – Elvina (South) Wharf – ferry – Church Point. **Distance :** 13.2 km

- 33. Church Point – ferry – Halls Wharf (Lovett Bay) – Towlers Bay Track – West Head Road - Bairne Track – Towlers Bay Lookout – Towlers Bay foreshore – Halls Wharf – ferry - Church Point. **Distance :** 12.6 km

- 34. Church Point – ferry – Elvina Bay (North) Wharf – Lovett Bay foreshores – Towlers Bay foreshores – Towlers Bay Lookout – Bairne Track – Bonnie Doon Wharf (The Basin) - ferry – Palm Beach. **Distance:** 10.5 km

- 35. Palm Beach (Pittwater) Wharf – ferry – The Basin Wharf – The Basin Track – West Head Road – Resolute Track – Great Mackerel Beach – wharf – Palm Beach. **Distance :** 8.2 km

- 36. Palm Beach (Pittwater) Wharf – ferry – Bonnie Doon Wharf (The Basin) – Bairne Track - Towlers Bay Lookout – Towlers Bay foreshore – Halls Wharf (Lovett Bay) – ferry - Church Point. **Distance :** 7 km

Native plant species mentioned in this book

Common name	Species
Black-eyed Susan	*Tetratheca* spp.
Black she-oak	*Allocasuarina littoralis*
Blady grass	*Imperata cylindrica*
Blueberry ash	*Elaeocarpus reticulatus*
Blue dampiera	*Dampiera stricta*
Bottlebrush	*Callistemon* spp.
Bracken fern	*Pteridium esculentum* or *Calochlaena dubia*
Broad-leaf cone-seed	*Conospermum longifolium*
Broad-leaf geebung	*Persoonia levis*
Brush muttonwood	*Rapanea howittiana*
Burrawang	*Macrozamia communis*
Cabbage tree palm	*Livistona australis*
Christmas bell	*Blandfordia nobilis*
Christmas bush	*Ceratopetalum gummiferum*
Coachwood	*Ceratopetalum apetalum*
Coast banksia	*Banksia integrifolia*
Conesticks	*Petrophile* spp.
Coral fern	*Gleichenia dicarpa*
Coral heath	*Epacris microphylla*
Crinkle bush	*Lomatia silaifolia*
Dagger hakea	*Hakea teretifolia*
Darwinia	*Darwinia fascicularis*
Dog rose	*Bauera rubioides*
Drumsticks	*Isopogon* spp.
Dwarf apple	*Angophora hispida*
Eggs and bacon	*Dillwynia retorta*
Flaky-barked tea-tree	*Leptospermum trinervium*
Flannel flower	*Actinotus helianthi*
Forest oak	*Allocasuarina torulosa*
Fuchsia heath	*Epacris longiflora*
Geebung	*Persoonia* spp.
Grass tree	*Xanthorrhoea* spp.
Grey grevillea	*Grevillea buxifolia*
Grey gum	*Eucalyptus punctata*
Grey ironbark	*Eucalyptus paniculata*
Grey mangrove	*Avicennia marina*
Grey myrtle	*Backhousia myrtifolia*
Grey spider flower	*Grevillea buxifolia*
Guinea flower	*Hibbertia* spp.
Heath banksia	*Banksia ericifolia*
Honey flower	*Lambertia formosa*
Hop goodenia	*Goodenia ovata*
Kanooka	*Tristaniopsis laurina*
Large-leaf bush-pea	*Pultenaea daphnoides*
Lillypilly	*Acmena smithii*
Long-leaf cone-seeds	*Conospermum longifolium*
Long-leaved lomatia	*Lomatia myricoides*
Mint bush	*Prostanthera* spp.
Narrow-leaf geebung	*Persoonia linearis*
Native broom	*Viminaria juncea*
Native holly	*Oxylobium ilicifolium*
Native rose	*Boronia serrulata*
Needle-bush	*Hakea* spp.
Old man banksia	*Banksia serrata*
Phyllota	*Phyllota phylicoides*
Pine-leaf geebung	*Persoonia pinifolia*
Pink grevillea	*Grevillea sericea*
Pink spider flower	*Grevillea sericea*
Pink swamp heath	*Sprengelia incarnata*
Pink tea-tree	*Leptospermum squarrosum*
Pinnate boronia	*Boronia pinnata*
Pittosporum	*Pittosporum undulatum*
Port Jackson fig	*Ficus rubiginosa*
Prickly moses	*Acacia ulicifolia*
Prickly tea-tree	*Leptospermu juniperinum*
Purple iris	*Patersonia sericea*
Red bloodwood	*Eucalyptus gummifera*
Red grevillea	*Grevillea speciosa*
Red spider flower	*Grevillea speciosa*
Red styphelia	*Styphelia tubiflora*
River mangrove	*Aegiceras corniculatum*
Rock-felt fern	*Pyrrosia rupestris*

Rough-barked apple *Angophora floribunda*
Scentless rosewood *Synoum glandulosum*
Scribbly gum *Eucalyptus haemastoma* or *E. racemosa*
She-oak *Allocasuarina* spp.
Silver banksia *Banksia marginata*
Smooth-barked angophora *Angophora costata*
Smooth-barked apple *Angophora costata*
Soft boronia *Boronia mollis*
Spider flower *Grevillea* spp.
Spine-leaved banksia *Banksia spinulosa*
Spotted gum *Eucalyptus maculata*
Stringybark *Eucalyptus* spp
Sundew *Drosera spathulata*
Swamp banksia *Banksia robur*
Swamp oak *Casuarina glauca*
Sweet sarsaparilla *Smilax glyciphylla*
Sweet-scented wattle *Acacia suaveolens*
Sydney boronia *Boronia ledifolia*
Sydney peppermint *Eucalyptus piperita*
Tall saw-sedge *Gahnia clarkei*
Tea-tree *Leptospermum* spp.
Turpentine *Syncarpia glomulifera*
Waratah *Telopea speciosissima*
Water gum *Tristania neriifolia*
Water vine *Cissus hypoglauca*
Wattle *Acacia* spp.
Waxflower *Eriostemon australasius*
Woollsia *Woollsia pungens*
Woolly star-hair *Astrotricha floccosa*
Yellow bloodwood *Eucalyptus eximia*